Guide to **IFC**98

Sarah Lupton

RIBA Publications

Biography

Sarah Lupton MA, DipArch, LLM, FCIArb, RIBA is a partner in Lupton Stellakis and lecturer in Architectural Practice and Management at the Welsh School of Architecture, Cardiff University. She is the RIBA deputy advisor on Adjudication, Arbitration and Practice Management, the RIBA examination moderator for professional practice, and represents the Consultants' College on the JCT working party for the Contracts (Rights of Third Parties) Act 1999. She lectures on courses and seminars run by the Royal Institute of Architects in Wales, the RIBA London, Eastern, and South Eastern Regions and University College London on a variety of subjects related to construction contract law. Sarah Lupton is author of *Guide to JCT98, Guide to MW98, Architect's Guide to Arbitration, Architect's Guide to Adjudication* and is editor of the seventh edition of the *Architect's Job Book*, also published by RIBA Publications.

First published 2001
Reprinted 2001

© Sarah Lupton 2001

Published by RIBA Companies Ltd, which trades under the name of
RIBA Publications, 1-3 Dufferin Street, London EC1Y 8NA

ISBN 1 85946 047 X

Series Title: **Guides to the standard forms of construction contract**
Series Editor: **Stanley Cox**

Publisher: **Mark Lane**
Designed by: **Jacqueline Stead**
Project Management and Editing by: **Elizabeth Davison**
Printed and Bound by: **The Periodical Press**
Typesetting: **Godfrey Lang Ltd**

While every effort has been made to check the accuracy of the
information given in this book, readers should always make their
own checks. Neither the author nor the publisher accepts any
responsibility for mis-statements made in it or misunderstanding
arising from it

Foreword

It gives me considerable pleasure to write this foreword. First, because the *Guide to IFC98* completes the trilogy of practical references from Sarah Lupton on the three main JCT lump sum forms of contract used for traditional procurement today. Sarah was my successor at The Welsh School of Architecture and she brings to this Guide experience of a practising architect, the communication skills of a distinguished academic, and a legally trained mind. She has kept the format and structure adopted for her earlier successful works on MW98 and JCT98.

Secondly, I admit to a rather nostalgic interest in the subject. When the IFC first appeared, I was fortunate to be invited to write the *Architect's Guide to IFC84*. RIBA Publications were ahead of the field with a book on the Intermediate Form, and this rather reflected the initiative and pioneering attitude of RIBA at the time, and in particular the work of its Contracts Committee and the then Legal Adviser, Robert Johnstone.

The IFC is deservedly popular with architects, for they played a great part in shaping it. When the Standard Form of Building Contract was introduced as a 1980 edition, there was considerable criticism of it by architects at regional appraisal meetings. Some regarded it as too heavyweight for their projects, and many felt that there should be an alternative for middle-range jobs. Some suggested that this might be a revised and shortened version of the earlier, simpler to administer JCT63; others argued that it could be an expanded Minor Works 80. Responding to such pressures, the then Joint Contracts Tribunal established a working party in October 1981 and within 12 months, drafting of a new Intermediate Form was sufficiently advanced to allow wide consultation to take place. The contract which eventually emerged was the result of consensus, but was very much what architects felt they needed.

IFC84 was different, even innovative. It came in only one version, with built-in adaptability, a logical section-headed format, facility for easy cross-referencing, and written in plain language. Provisions such as those for naming sub-contractors, deferment of possession, and instructions following failure of work, were not found in other JCT forms at the time. Procedural rules were kept to a minimum and it was relatively simple to administer. Interestingly it has not featured much in case law!

IFC98 is a consolidated version of IFC84 and incorporates all those Amendments which appeared over fourteen years. Whilst many of the original attractive IFC features were later introduced into other JCT forms of contract, conversely some of the heavier provisions introduced as Amendments into JCT80 also appeared in the Intermediate Form. As a result IFC98 is no longer as short, or easy to understand and administer, as its predecessor. That is one good reason why you need to discard the old IFC guides and read this up-to-date practical reference. I learned a lot about using the current Intermediate Form from reading this guide!

Stanley Hall Cox MBE FRIBA FCIArb
Series Editor

Contents

Contents

About the guide

This guide is intended to act as a practical aid for those interested in the operation and administration of IFC98. It explains the provisions of the contract and how they might work in practice. It assumes only a general knowledge of JCT traditional forms of contract, and can serve as an introduction to IFC98 for the student and the recently qualified construction professional, as well as providing a 'desktop' reference for the busy contract administrator.

The commentary is set out in a sequence close to that of the clauses of the form, although adjustments are made where it is more logical to group like topics together.

The guide gives a broad outline of the form and the reasons why it might be selected. It describes the various documents that may form part of the contract package at the time it is entered into. It examines the contractor's obligations with respect to quality and standard of work, including possible design obligations. It deals with matters of programming and extensions of time and examines the mechanisms available, including the powers of the administrator, to monitor and control the standard of work. It analyses the procedures for adjusting the contract sum and the process of certification. It also deals with matters of insurance, determination and dispute resolution. Much of the information is in tabular or diagrammatic form to enable quick and clear access to information.

The appendices to this guide include extracts from some relevant JCT Practice Notes and from the notes that accompanied amendments to JCT forms. Some key court cases are referred to, each of which has a paragraph 'case note'. The guide is not intended to be a comprehensive legal commentary, and should difficult points arise the reader should consult one of the well-known legal texts or take legal advice.

The guide does not include detailed advice on completing the contract administration forms produced by the RIBA for use with the JCT Standard Form of Building Contract. For this the reader should refer to the relevant *Architect's Guide to the Contract Administration Forms: IFC98* published by RIBA Publications. This also includes checklists of action to be taken by the contract administrator both pre-contract and during work on site.

The author would like to thank Mark Lane of RIBA Publications and would particularly like to express her gratitude to Stanley Cox, the Series Editor, not only for his invaluable help, but also for his enthusiasm and encouragement throughout the writing of this book.

1.1   The Intermediate Form of Building Contract (IFC98) is published in only one version, for use by both private clients and local authorities. At the time of writing the latest reprint of IFC98 incorporates Amendments 1 (1999) and 2 (2000), with Amendment 3 (2001) published separately. When first published, IFC98 was in effect a re-publication of IFC84 with Amendments 1-12 incorporated, together with some further minor corrections and alterations. Amendment 12 introduced the changes required by the Housing Grants, Construction and Regeneration Act 1996, together with other changes largely derived from recommendations of Sir Michael Latham, all of which are now incorporated into IFC98.

1.2   There are several ancillary documents published for use with IFC98. A complete suite of forms is published for use with named sub-contractors. This comprises a form of tender and agreement (NAM/T: three sections), sub-contract conditions (NAM/SC), and sub-contract fluctuation rules (NAM/SC/FR), all published by the JCT. There is also a warranty for use between the employer and the named sub-contractor, the Employer/Specialist Agreement (ESA/1), which is published by RIBA/CASEC. There is a Sectional Completion Supplement which allows for phased possession and completion (IFC/SCS). Fluctuation Clauses and Formula Rules, and the JCT Adjudication Agreement, to be used when appointing an adjudicator, are also published separately. The JCT collateral warranties to a funder (MCWa/F) and purchaser or tenant (MCWa/P&T) may be used with IFC98.

Key features

1.3   IFC98 is a traditional lump sum contract, where the contractor is required to carry out the work described briefly in the first recital and shown in the contract documents for the Contract Sum entered in Article 2, and to complete the work by the date entered in the Appendix. The contract makes reference throughout to an 'Architect/Contract Administrator' who is named in Article 3, and given various powers and duties under the contract, including the obligation to supply the contractor with all information reasonably necessary for carrying out the Works. The contract contains provisions for varying the work, and adjusting the Contract Sum and the Date for Completion on the occurrence of certain events.

1.4   The quality and quantity of work to be carried out is that shown in the contract documents, therefore it is extremely important that the description of the Works is accurate and comprehensive. The contract allows for a variety of combinations of contract documents, which may include drawings, a Bill of Quantities, specifications, schedules of work and schedules of rates. At tender stage the contractor prices either a Bill of Quantities, or specification, or schedule of works, or submits a Contract Sum analysis or schedule of rates as required. IFC98 is a lump sum contract which means the Works shown in the contract documents must be carried out for the Contract Sum 'or such other sum as shall become payable hereunder at the times and in the manner specified in the Conditions' (Article 2).

1.5   Payment to the contractor is made at monthly intervals following the issue of architect's certificates. In general terms, the certificates will reflect the amount of work that has been properly completed up to the point of valuation in accordance with the terms of the contract. All the requirements of the Housing Grants, Construction and Regeneration Act 1996 regarding payment and notices are incorporated into IFC98, and the provisions regarding contractor's price statements, contractor's applications for payment, listed items, activity schedules and advance payments which were introduced into JCT98 alongside the 1996 Act's requirements, have also been stepped down into IFC98.

1.6   As with other JCT traditional forms the contractor may sub-contract the work to domestic sub-contractors with the written approval of the architect. The key distinguishing feature of IFC98, however, is the provisions allowing for the naming of sub-contractors. Under these provisions the contractor can be required to sub-contract to a specific contractor or supplier, who may be either named in the tender documents, or after the contract is let in an instruction for the expenditure of a provisional sum. This allows the employer a great deal of flexibility and a high degree of control over who carries out the work and the terms on which it is done. The employer, however, takes some risk where, for example, the contractor may have the right to an extension of time for delays resulting from instructions relating to named sub-contractors (see figure 1).

1.7   An additional flexibility is that the named sub-contractor provisions also allow for the sub-contractors to carry out design, the only mechanism within IFC98 by which a design element can be included in the Works to be carried out. If this is required it is essential that a direct warranty is arranged between the employer and the sub-contractor, as IFC98 excludes any liability of the contractor for the design by named sub-contractors.

Deciding on IFC98

1.8   JCT Practice Note 6 (series 2) *Deciding on the Appropriate Form of JCT Main Contract* advises that IFC98 is suitable where the contract period is not more than 12 months and the Contract Sum is not more than £375,000 (2001 prices). In addition, it is stated on the back of the form to be suitable 'where the proposed building works are –

1. of a simple content involving the normally recognised basic trades and skills of the industry; and
2. without any building service installations of a complex nature, or other specialised work of a complex nature; and
3. adequately specified, or specified and billed, as appropriate prior to the invitation of tenders'.

Figure 1 Distribution of risk with named sub-contractors

	Revision to contract sum?	Extension of time?	Loss and/or expense?
named sub-contractor's progress causes delay	no	no	no
delay in issuing instructions dealing with naming	no	yes (2.4.7.2)	yes (4.12.1.2)
compliance with instructions under 3.3.1(a), (b), (c)	yes, 3.6	yes (2.4.5)	yes (4.12.7)
compliance with instructions under 3.3.2(a) (provision sum)	yes, 3.6	yes (2.4.5)	yes (4.12.7)
following determination, compliance with instructions naming replacement sub-contractor under clause 3.3.3(a)			
if original sub-contractor named in contractor documents (cl 3.3.4(a))	yes	yes (2.4.5)	no
if original sub-contractor named in instruction (cl 3.3.5(a)	yes	yes (2.4.5)	yes (4.12.7)
following determination, compliance with instructions under 3.3.3(b) or (c)	yes	yes (2.4.5)	yes (4.12.7)
following determination, where contractor has not complied with 3.3.3	down only	no	no

1.9 In practice, IFC98 is suitable for a very wide range of project types and size, and certainly the limits of 12 months and £375,000 should be treated as only an approximate guide. There are, however, some situations where it would not be suitable. Care should be taken where the client is a consumer, particularly where carrying out work to a dwelling which the client intends to occupy. The Housing Grants, Construction and Regeneration Act 1996 does not apply in this situation, and some of the 1998 revisions may be considered unfair under the Unfair Terms in Consumer Contract Regulations 1994 (SI 1994/3159), if not individually negotiated with the client prior to entering into the contract. The contract may also be unsuitable where the quantity or type of work is largely unknown, or where the work is extremely complex.

1.10   In particular, IFC98 would not be suitable where it is anticipated that the main contractor will be required to carry out design. There is no equivalent to the JCT98 Contractor's Design Portion Supplement for use with IFC98, and the contract contains no clauses to deal with the implications of performance specification. If main contractor design is needed, then another form should be considered.

1.11   On the other hand, the provisions for naming of sub-contractors may be particularly useful where design is required by known specialist firms. This is a common requirement in practice with specialist design items such as mechanical services or pre-fabricated elements. It is possible to enter into the warranty ESA/1 with the specialist at an early stage, so that the design can be integrated into the project as a whole, with the safeguard to the specialist that if it is not awarded the sub-contract it will be paid for any design work carried out under the warranty agreement. Even where design is not an issue, there may be circumstances in which the employer wishes part of the work to be carried out by a particular firm.

1.12   There are however, some risks to the employer in naming sub-contractors, although these are not so extensive as with the nominated sub-contractor provisions in JCT98. They are domestic sub-contractors, and generally the contractor takes full responsibility for the performance of named sub-contractors, including for their progress on site, except that delays in the naming process, eg in issuing instructions naming the sub-contractor, would be the employer's risk, as would delays resulting from a determination of the sub-contract due to the sub-contractor's default. The main contractor is also entirely responsible for payment – there is no requirement to name amounts due to named sub-contractors in certificates, nor any obligation on the employer to pay the sub-contractor direct should the contractor fail to do so. In addition, the procedures associated with naming the sub-contractor are slightly simpler than those for nomination, although they are still by no means straightforward. On balance, the naming provisions have proved popular with construction professionals, and are often one of the main reasons for selecting this form.

1.13   The form is laid out under 'section headed' format, which by and large groups the clauses in a logical fashion, and makes it much easier for the user to find their way around than is the case with JCT98. Rather curiously, the named sub-contract conditions are not laid out in this manner, so moving between documents is unnecessarily confusing. This is compounded when trying to work with the Sectional Completion Supplement.

1.14   It is perhaps a pity that all the amendments introduced into JCT80 by
        Amendment 18 were stepped down into IFC84, as the complexity of some of
        these is out of proportion to the form as a whole, and has tipped the balance
        of risk in favour of the contractor. In addition, some of the other clauses seem
        overly complex for a form of this scale, for example the determination
        provisions, perhaps again as a result of amendments to the larger form being
        stepped down without modification.

1.15   Overall, though, it is a much more workable document than JCT98 for relatively
        straightforward projects of the scale envisaged, and avoids many of the gaps and
        pitfalls of MW98. It is therefore a popular choice for medium to large scale projects.

2 Documents

2.1 Documents are a key factor in the success of every building project, and traditional projects such as those under IFC98 depend on comprehensive and accurate information at tender stage, with any supplementary information being provided in adequate time and to an agreed and workable programme. Inadequate and late information is more than likely to result in an escalating budget and delays to the completion date.

2.2 IFC98 makes reference to various documents, not all of which are termed 'Contract Documents'. These other documents may nevertheless be included in the tender documents, or form part of the contract by incorporation by reference, or in some other way be of contractual effect. The recitals refer to drawings, Bills of Quantities, specification, schedules of work, NAM/T, 'Numbered Documents', a Contract Sum analysis, a schedule of rates, a priced activity schedule, and an information release schedule. The conditions refer to numerous further documents including, an 'advance payment bond', the Health and Safety Plan, and the Joint Fire Code. Figure 2 indicates some of the possible combinations of documents that may make up the contract package for IFC98.

Figure 2 Possible combinations of documents

Second recital:		
	Alternative A	**Alternative B**
Drawings	CD	CD
Priced Specification/		
Schedules of work/Bill of Quantities	CD	
Un-priced specification		CD
Contract Sum analysis/Schedule of rates		R
Priced activity schedule	R	R
Information release schedule	R	R
Listed items and related bond	cl 4.2.1	cl 4.2.1
Advance payment bond	cl 4.2 (d)/Appendix	cl 4.2 (d)/Appendix
Health and safety plan	cl 5.7.2	cl 5.7.2
Collateral warranty	cl 3.1.1/3.16	cl 3.1.1/3.16
KEY		
CD = a 'Contract Document'		
R = referred to in the recitals		
cl = referred to in the Conditions		

Documents referred to in the recitals

2.3   The 'Contract Documents' are defined under the second recital. There are two alternatives. Under alternative A the contract documents will consist of:
- either a priced specification, or priced schedules of work, or a priced Bill of Quantities (signed by the parties);
- the contract drawings (signed by the parties);
- NAM/T with Sections I and II completed and signed, for all sub-contractors named in the contract documents;
- the Agreement (completed and signed by the parties) and Conditions, both found in the printed form.

2.4   Under alternative B the contract documents will consist of:
- a specification (unpriced and signed by the parties);
- the contract drawings (unpriced and signed by the parties);
- NAM/T with Sections I and II completed and signed, for all sub-contractors named in the contract documents;
- the Agreement (completed and signed by the parties) and Conditions, both found in the printed form.

2.5   In both cases the contract documents will include the printed form, contract drawings and NAM/T Sections I and II. The essential difference between the two sets of documents is that in alternative A the contractor is given the document to be priced (specification, Bills or schedules), which subsequently becomes a contract document, whereas with alternative B the contractor simply states a sum and provides a breakdown in the form of a contract sum analysis, or provides a schedule of rates, neither of which is termed a 'contract document'. A priced activity schedule can be used with either option but is also not termed a 'contract document'. It should be noted, however, that if used, the contract sum analysis, schedule of rates or activity schedule will have a contractual effect.

2.6   The second recital (alternative A) refers to the contractor having 'priced the Specification/priced the Schedules of Work/priced the Bills of Quantities', and two of these should be deleted as appropriate. Bills of Quantities would normally be used on larger or more complex projects, and would be prepared by the quantity surveyor according to the rules in the Standard Method of Measurement 7th edition (SMM7) **(cl 1.5)**. Bills are particularly helpful when it comes to the accurate comparison of tender figures, and valuation of variations. Schedules of work are often a useful alternative for smaller projects, and can be arranged in any appropriate format. This is frequently by work sections or by trade, although a 'room by room' basis is sometimes used in refurbishment projects. Drawings plus specification alone would normally only be used on simple projects where few variations are anticipated. Whichever document is used, it should be noted that if materials or goods are to be paid for prior to delivery on site a list of these must be annexed (cl 4.2.1(c)).

2.7　Even where either Bills or schedules of work are used, a specification would normally be prepared by the architect, and in some cases all three types of document may form part of the project information. If the specification and/or schedules are to form part of the tender package alongside the Bills, it is best to organise these as one document, ie as numbered sections of the Bills. Alternatively a schedule may form one of the numbered drawings. If a document sent out to tender is not referred to accurately in the first recital, there could be room for doubt later as to whether it forms part of the binding agreement between the parties.

2.8　Under the second recital (alternative B) the contract sum analysis or schedule of rates could be in any form required, eg the contractor could even be asked to prepare a full Bill of Quantities, although this would be unlikely on smaller projects. The document will be used for preparing valuations and assessing the value of variations, so it would be wise to set out what format would be acceptable in the tender documents. (Guidance on the use of this document can be found in JCT Practice Note 23 *A Contract Sum Analysis.*) A schedule of rates on its own is unlikely to be a very useful document. It is important to note that although the contract sum analysis is not a contract document, it becomes a 'priced document' for the purposes of clause 3.7.2 (valuing of variations) and to that extent the employer is bound by its terms.

2.9　The Articles of Agreement and the Appendix must be completed very carefully (for guidance on completing the form see *Architect's Guide to the Contract Administration Forms: IFC98*). The Articles of Agreement contain the attestation that must be signed by both parties and witnessed, and special procedural steps must be taken if the contract is to be executed as a deed.

2.10　The 'Contract Drawings' are listed under the first recital. These should all be identified precisely, including revision numbers, etc. The list may be annexed if long, but if so the list must be clearly identified. Note that in IFC98 there is no reference to who prepared the drawings. The contract requires that the parties sign all the drawings. For good practice in preparation and coordination of specification, drawings and Bills of Quantities, see current relevant publications on Coordinated Project Information (CPI).

Activity schedule

2.11   The second recital refers to a priced activity schedule, which may be required under both alternative A and alternative B. If an activity schedule is not required the provision must be deleted. The activity schedule is defined in clause 8.3 as a 'schedule of activities', which should be attached to the Appendix to the form. The schedule is prepared and priced by the contractor and provided prior to the contract being executed. An example of a priced activity schedule was included in the guidance notes to JCT80 Amendment 18, and it is very similar to a schedule of work *(see Appendix A)*. Each activity is priced, and the sum of those prices must equal the contract sum with certain exclusions, namely provisional and prime cost sums and related contractor's profit, and the value of work for which approximate quantities have been included in the contract documents.

2.12   This is rather curious provision to find in the form, where it is already anticipated that there will be a priced document which, with the possible exception of the priced specification under alternative A and the schedule of rates under alternative B, is likely to have the price broken down in considerable detail. The function of the activity schedule is to ascertain the value of work properly executed for certification purposes, and if included, the breakdown on the activity schedule will be used rather than that given in any other priced document. It will certainly give clear information as to how much value the contractor attaches to each activity, and ensure that the certificates reflect this. However it would be possible for contractors to 'front load' the activity schedule in order to improve their cash flow in the early stages, as there is no obligation for the schedule to reflect the prices given in the Bills. Where it is to be used, the architect should therefore be alert to any problems of this sort, which should be dealt with at tender stage before the contract is formed. It should be noted that even if an activity schedule is used, the priced document will remain the basis for the valuation of variations.

Information release schedule

2.13   The information release schedule is referred to in the fourth recital. It is an optional provision (the recital is deleted if the schedule is not provided). The schedule should state 'what information the Architect/Contract Administrator will release and the time of that release'. If used, the schedule is prepared by the architect and sent out with the tender documents. The schedule does not need to list all the information that will be provided, but could, eg, list key drawings.

2.14   If the information release schedule is used, then the information shown in it must be supplied at the dates indicated (cl 1.7.1). Failure to provide the information is a ground for an extension of time (cl 2.4.7) and a 'matter' which may give rise to a direct loss and/or expense claim (cl 4.12.1). With respect to information not shown on the schedule, or where a schedule is not used, the architect is under an obligation to provide 'such further drawings and details as are reasonably necessary' (cl 1.7.2) either in sufficient time to allow the contractor to complete by the Date for Completion, or, if the contractor appears unlikely to complete by this date, at a date when 'having regard to the progress of the works' it is reasonably necessary for the contractor to receive the information. Failure to provide this information is again a ground for an extension of time (cl 2.4.7) and a 'matter' which may give rise to a direct loss and/or expense claim (cl 4.12.1).

2.15   The advantage of using the information release schedule is that it gives the architect the opportunity to prepare a realistic programme of drawing production, so that the contractor has a clear picture regarding information provision before submitting a tender. The contractor will not be able subsequently to request information by dates earlier than those shown, thus the schedule will prevent the manoeuvring which sometimes occurs where lists of information are requested at the start of a project in the hope of setting the scene for a later claim. Used carefully, the information release schedule can therefore be a very effective management tool, provided the programme set out does not contain too much 'wishful thinking' on the part of the consultant team! If the contractor feels the schedule is awkward in term of its planned operations, then adjustment could be negotiated at tender stage. The contract contains no provisions allowing the architect unilaterally to adjust the programme, for example where an extension of time is granted or the contractor is running behind programme. As it would be to both parties' advantage for the document to remain a realistic representation of when information is required, it may be sensible to make allowance in the tender documents for updates to be negotiated, perhaps at monthly progress meetings.

IFC98 Supplements

2.16   The Sectional Completion Supplement is a schedule of amendments which adapts IFC98 to allow for phased completion (and commencement) of the Works. If phasing is planned, it is essential to make this clear in the tender documents and to incorporate the Supplement from the outset. The Supplement amends the first recital to refer to the division of the Works into 'Sections', and these must be clearly identified in the contract documents, with the divisions being selected on a workable basis. The guidance notes incorporated in the Supplement recommend that any part of the Works that is common to several sections (an example might be mechanical services) is identified as a separate section.

2.17 If the Supplement is used, the Conditions are modified so that they operate independently with respect to each section (in most instances the term 'Works' is either replaced by 'Section' or 'Sections' or has a reference to 'Sections' added). For example, the contractor must notify the architect of delays to any section, and there is provision for issuing non-completion certificates and fixing new completion dates for each section as appropriate. A separate certificate of Practical Completion and certificate of making good defects is required for each section, but only one final certificate for the Works. There are implications for liquidated damages and retention.

Other documents
Bonds
2.18 IFC98 refers to two bonds: (1) an advance payment bond, and (2) a bond in respect of payment for off-site materials and/or goods (note the provisions for advance payments do not apply where the employer is a local authority). Where required, the contractor must arrange bonds, and as both are optional it must be made clear to the contractor at tender stage if either will be required. The former is normally required where an advance payment is to be made to the contractor under clause 4.2(b), and the latter where it has been agreed that certain materials or goods will be paid for in advance of them being brought on site (cl 4.2.1). Terms for each of the bonds have been agreed between the British Bankers Association and JCT Ltd and are included in the form as Annex 1 to the Appendix. If any other terms are preferred, or if any other type of bond is required, eg a performance bond, then the fifth recital makes it clear that these must have been given to the contractor before the contract is entered into. In practice the alternative terms should be sent out with the tender documents, so that the contractor can include them in the tender figure. If the contractor proposes alternative terms, then these should be forwarded to the client for discussion with the client's lawyers. Architects do not normally have sufficient knowledge to advise the client on the terms of bonds.

Collateral warranty
2.19 Although not referred to in the contract, the JCT publishes a Form of Agreement for Collateral Warranty in two versions. The first, MCWa/F, is for use where a contractor is required to give a warranty to a company providing finance for the building works. The second, MCWa/P&T is for use where a contractor is required to give a warranty to a purchaser or tenant of the building works. Both forms introduce an 'enabling clause' (which can be inserted as either clause 3.1.1 or clause 3.16) into the contract whereby the employer may

require the contractor to enter into the warranty. The requirement must, of course, be made clear at the time of tender.

Health and safety documents

2.20 Clause 5.7.1 to 5.7.4 will be applicable where in the Appendix it is shown that all the Construction (Design and Management) Regulations 1994 (SI 1994/3140) ('the CDM Regulations') apply to the particular contract, and this will usually be the case with IFC98. The clause specifically refers to the Health and Safety Plan, and the Health and Safety File.

2.21 The Health and Safety Plan is not a contract document under IFC98, and the recitals make no mention of it having been prepared and given to the contractor at the time of tender. Nevertheless, it is a statutory obligation for the employer to have had one prepared and passed to the principal contractor under regulation 15 of the CDM Regulations. As it may have pricing and timing implications a 'pre-tender' Health and Safety Plan is usually sent out with the tender documents. Before construction work can begin the contractor must have developed the plan to comply with regulation 15(4). As the contract contains no sanctions for its non-provision, other than the rather draconian measure of determination under clause 7.2.1(e), it may be wise to insist on this before the contractor is appointed. If this is to be a pre-condition for appointment, this should be made clear in the tender documents.

2.22 The Health and Safety File is principally a matter for the planning supervisor who will compile it, but there is a requirement on the contractor to provide information for this File, and to ensure any sub-contractor also complies. The architect, when certifying Practical Completion, must make sure that the contractor has 'complied sufficiently' with this requirement before issuing the Certificate.

Use of documents
Interpretation, definitions

2.23 Clause 8.3 of the Conditions sets out definitions of terms that are used throughout the contract. Some further and more detailed definitions are embodied in the text of clauses, for example 'All Risks Insurance' and 'Joint Names Policy' are defined at the beginning of clause 6.3.2. Some items were first introduced by Amendment 12 to IFC84. Those regarding notices and periods of time were required by the Housing Grants, Construction and Regeneration Act 1996 and re-state its requirements relating to the serving of notices and the calculation of periods of days (cl 1.13–1.14). It should be noted that the contract sets out specific requirements for notices in some situations, for example determination (cl 7.1) which would override the general provision in clause 1.13.

2.24   Clause 1.15 states that the law of the contract will be English law. This would apply even if the contract was signed, or the work was carried out, in another jurisdiction.

2.25   Clause 16 refers to the supplemental provisions for electronic data interchange (EDI) in Annex 2 to the Conditions. If the employer wishes the supplemental provisions to apply then this must be made clear at tender stage, and the parties must enter into an electronic data interchange agreement, prior to entering into the main contract.

2.26   Clause 1.17 was introduced through Amendment 2 in response to the coming into effect of the Contracts (Rights of Third Parties) Act 1999. In broad terms this Act created rights for persons not a party to a contract to bring an action for breach of a contract, where that contract expressly gave a benefit, or purported to give a benefit to that person. This clause prevents any such claims being brought by making it clear that the contract confers no rights on third parties.

Priority of contract documents

2.27   Clause 1.3 states '...Nothing contained in the Contract Bills shall override or modify the application or interpretation of that which is contained in the Articles of Agreement, the Conditions or the Appendix ...'. If this clause was not included the position under common law would be the reverse, in other words anything specifically agreed and included in a document would normally override any standard provisions in a printed form.

2.28   If the parties wish to agree to any special terms that differ in any way from the printed conditions, then the amendments will need to be made to the form. This could be done either through amending the clauses themselves, or inserting an additional article referring to the special terms, which should be appended to the form. The article could take a similar form to that used by the JCT to incorporate separately published amendments. However, attempting to amend standard forms is very unwise without expert advice as the consequential effects are difficult to predict. Deleting clause 1.3 could be particularly unwise as it may have unintended effects on other parts of the contract.

2.29   With regard to quality and quantity of work, clause 1.2 sets out details as to priority between documents (other than the form itself). If Bills are used, the quality and quantity of work shown is the work which is to be provided, even if the other documents (eg drawings) give different information. Where there are no Bills, but the specification or schedules of work give quantities, the quantity and quality shown would override any conflicting information in other documents. Where there are no Bills, and no quantities given in the specification/schedules, the quality and quantity of the work 'shall be deemed to be that in the Contract Documents taken together, provided that if work stated or shown on the Contract Drawings is inconsistent with the description, if any, of that work in the Specification/Schedules of Work then that which is stated or shown on the drawings shall prevail for the purposes of this clause'. In broad terms the drawing will take precedence over other documents if those documents do not contain any quantities.

Inconsistencies, errors or omissions

2.30   Clause 1.4 requires the architect to issue instructions in regard to: inconsistencies, errors or omissions in or between the contract documents, and in or between any further information issued to the contractor; any errors or omissions in the named sub-contractor particulars, or any departure from the agreed method for preparing any Bills of Quantities.

2.31   The contractor is not under any express obligation to point out any inconsistencies, errors or omissions, although generally it would be in the contractor's interest to do so, and a certain amount of vigilance could be expected under the normal duty to use reasonable skill and care. Nevertheless it is the responsibility of the consultants to identify such problems, and the architect's duty to issue the necessary instructions. If the instruction results in a change to the quantity or quality of the work, or to any restrictions imposed by the employer, then the contractor may be entitled to additional payment. In addition, the instruction may give rise to a claim for an extension of time (cl 2.4.5) and for loss and/or expense (cl 4.12.7).

Custody and control of documents

2.32   The contract documents remain in the custody of the employer, and must be available for inspection by the contractor at all reasonable times (cl 1.6). The architect should retain a copy for reference throughout the life of the contract. The contractor must be provided with one certified copy of the contract (including all 'the Contract Documents') and two further copies of the contract drawings and the specification/schedules of work/contract Bills (cl 1.6). It does not state when this should be done, but it would be good practice to arrange for the contractor to have the copies prior to work starting on site. Although it is frequently done in practice, there is no need to sign two copies of the contract. It is easy to make minor mistakes when filling out two copies of the form, and it is safer to have one definitive set of contract documents, with certified copies made as required.

2.33   The documents provided to the contractor (including the contract documents and any further information) must not be used for any purpose other than the Works and the details of the rates or prices are not to be divulged by the quantity surveyor or the architect (cl 1.8).

Sub-contract documents
Domestic sub-contracts

2.34   JCT Ltd does not currently publish a standard form for use with domestic sub-contracts, and although one is published by the Construction Confederation (IN/SC), there is no requirement under IFC98 that the main contractor should use this form. There are, however, restrictions on the terms that may be agreed. These are set out in clause 3.2.2 of IFC98, which requires that particular conditions relating to ownership of unfixed goods and materials, and the right to interest on unpaid amounts properly due to the sub-contractor, are included in all domestic sub-contracts. The sub-contract should also, of course, comply with the requirements of the Housing Grants, Construction and Regeneration Act 1996.

Named sub-contracts

2.35   JCT Ltd publish three documents for use with the named sub-contract provisions:
NAM/T: the form of tender and agreement, consisting of three parts;
NAM/SC: the conditions of contract between the main contractor and the named sub-contractor;
NAM/SC/FR: sub-contract fluctuation rules.
In addition, RIBA and CASEC jointly publish a warranty for use between the employer and the named sub-contractor (ESA/1).

2.36   If it is anticipated that named sub-contractors will be used then either the sub-contractor should be named in the Bills (or specification/schedules of work), or a provisional sum included as described under clause 3.3.2. The 'Sub-contract Documents' are defined as 'Tender and Agreement NAM/T, the Sub-contract Conditions and the Numbered Documents' (NAM/T Section III, article 1.1). The 'Numbered Documents' describe the 'Sub-contract Works' to be carried out. No detailed description is given of what the documents might be; in practice they are likely to comprise drawings, specification, Bills, schedules, etc, which will have been prepared by the architect. In some cases, where the named sub-contractor is involved in design, they will also include information prepared and submitted with the tender by the sub-contractor.

2.37   NAM/SC clause 1.3 sets out definitions of terms used in the sub-contract documents. Clause 2.2 states that in the event of any conflict, the terms of NAM/T would take precedence over terms in any of the other sub-contract documents. If any conflict arises between the terms of the main contract and the terms of the sub-contract documents then the latter will prevail. NAM/SC contains similar provisions to IFC98 regarding the correction of errors, etc. Where there are inconsistencies in or between the sub-contract documents, the contractor must issue directions to correct them (NAM/SC cl 2.4).

3 Obligations of the contractor

3.1 The contractor's paramount obligation is to 'carry out and complete the Works'. This obligation, which is stated in Article 1, reinforced in clause 1.1, and amplified in clause 1.2, is discussed in detail below. In addition, the contractor has important obligations in relation to progress and programming, discussed in chapter 4, and in regard to insurance matters, discussed in chapter 7. The contractor's obligations under IFC98 are summarised in figure 3.

Figure 3 Key obligations of the contractor

IFC98	Key obligations of the contractor
1.1	to carry out and complete the works in compliance with the contract documents
2.1	to begin the Works on the date of possession and to proceed regularly and diligently and complete on or before the Date for Completion
2.3	to notify the architect whenever it becomes reasonably apparent that the progress of the Works is being or is likely to be delayed
2.7	to pay the employer liquidated damages for failure to complete by the completion date
2.10	to make good defects scheduled by the architect at the end of the defects liability period
3.2.2	to ensure that any sub-contract includes specified conditions
3.3.1 & 3.3.2	to enter into sub-contracts with named persons
3.3.3	to advise architect of events which may lead to determination of employment of named persons and notify architect of any determination
3.4	to constantly keep upon the site a competent 'person-in-charge'
3.5.1	to forthwith comply with all instructions issued by the architect
3.9	to correct any setting out errors
3.11	to permit the employer to carry out work not forming part of the contract
4.5	to provide the architect with all documentation necessary for finalising the final account
5.1 & 5.7.2	to comply with all statutory requirements
6.1.1 & 6.1.2	to indemnify the employer and to take out effective insurance against injury to persons and property
6.3	to take out insurance against loss or damage to the Works as required
6.3FC	to comply with the Joint Fire Code where it applies

The Works

3.2 The Works that the contractor undertakes to carry out will be as briefly described in the first recital of IFC98, and as shown or described in the contract documents. It is therefore important that the entry in the first recital clearly identifies the nature and scope of the proposed work, and that the descriptions of the Works set out the required standards and quality of workmanship and materials fully and accurately. It should be noted that the contractor's obligation to carry out the Works extends to any changes made to those Works, as stipulated in Article 1, provided that these changes are made in accordance with the terms of the contract.

A design role?

3.3　IFC98 is essentially a work and materials contract, and makes no reference to the main contractor carrying out design. The only mechanism within the contract whereby some of the design responsibility can be allocated to a specialist company is through the use of the provisions for named sub-contractors. This allocation of design liability must of course be approved by the client before the named sub-contractor is approached, and preferably before the architect's appointment is agreed with the client, as otherwise the architect will remain fully responsible for all aspects of that design.

3.4　If any attempts are made to place a design obligation on the contractor using 'ad hoc' methods, eg through the use of clauses in the Bills or by the inclusion of a performance specification, it would depend on the particular circumstances whether or not this would be successful. Such methods should not be attempted without legal advice. Bearing in mind the result of the case of *Rotherham MBC v Frank Haslam Milan* it is clear that a court would be unwilling to place design responsibility on a contractor in situations where an architect has been employed to prepare full design and specifications.

Rotherham MBC v Frank Haslam Milan and MJ Gleeson (1996) 78 BLR 1 (CA)

A contract (JCT63) to construct a new office building involved the laying of fill on the site and the contractor, Gleeson, was given a detailed specification as to the material of the fill: '... hardcore shall be graded or uncrushed gravel, rock fill, crushed concrete or slag or natural sand or a combination of any of these'. The clause omitted to state the types of slag that could be used and the unweathered steel slag selected by the contractor proved to be expansive, causing cracking in the concrete ground floor slab (around £700,000 worth of damage). The Court of Appeal, overturning the decision of the Official Referee at first instance, found the contractor was not liable for the damage caused by the heave. The court stated, among other things, that the fact that the employer had engaged an architect to write the specification showed the employer did not intend to rely on the skill and expertise of the contractor in selecting the hardcore. If it had not been for these special circumstances, however, the court stated that it would have found the contractor strictly liable for providing a slag that was fit for the purpose for which it was to be used. It should be noted that *Rotherham* was concerned with an omission within the design information, and that the Court of Appeal made it clear that it was not the express terms of JCT63 that prevented the contractor assuming design liability, but the surrounding circumstances. If the parties had made it clear, eg through the specification or a note on a drawing, that they were relying on the contractor to select a suitable slag, then the outcome might have been different.

Standards and quality

3.5   Under clause 1.1 the contractor is obliged to carry out the Works in a proper and workmanlike manner, and in accordance with the contract documents and the Health and Safety Plan. Clause 1.2 discussed the position where there is a conflict between the quality stipulated in different contract documents. Although the architect must issue instructions to resolve the discrepancy, there could be doubt as to the contractor's original obligation as to quality, and hence whether or not the instruction would constitute a variation. Clause 1.2 therefore sets out a priority between documents. Where there is a description given in a Bill of Quantities, this would take precedence over any description contained in another document. Where no Bill of Quantities has been used, but quantities are set out in a specification or schedule of works, the quality of work should be as described in that document. In situations where Bills are not used and the specification or schedules do not include quantities, the quality is that 'in the contract documents taken together', except that any description given in drawings would override any set out in a specification or schedule of work.

3.6   In cases not covered by the above, and in cases where the documents must be 'taken together', a court would assess what, from the point of view of an objective bystander, appeared to be the intention of the parties at the time the contract was entered into. In practical terms, care should be taken to avoid duplicating descriptions of quality. The Committee for Coordinated Project Information (CCPI) recommendation is that the specification becomes the core document in terms of defining quality, and that the drawings, schedules and the Bills refer to clauses in the specification, and careful use of the CCPI system ought to avoid conflicting descriptions appearing in different documents.

3.7   If no standard is given at all then it is suggested that following the case of *Rotherham MBC v Frank Haslam Milan* discussed above, the contractor would be unlikely to be held liable to provide materials fit for their intended purpose, and the onus would be on the architect to be alert to possible errors or omissions in the description of the quality to be provided, and to issue appropriate instructions

3.8   Unlike JCT98, the obligation to provide materials in accordance with the contract is not qualified by the phrase 'so far as procurable'. This means that failure to supply would be a breach of contract by the contractor, even if the item was unavailable. However, if the item is of fundamental importance and unavailability brought the project to a standstill, it could be considered that the contract was frustrated, which would relieve the contractor of any obligation to continue with the Works. Normally, of course, the architect would issue an instruction changing the item, which would result in a variation. It would be an implied duty that the contractor should notify the architect before substituting any materials or goods, even where those specified are unobtainable.

3.9   The phrase 'where and to the extent that approval of the quality ... is a matter for the opinion of the Architect/the Contract Administrator, such quality and standards shall be to the reasonable satisfaction of the Architect/the Contract Administrator' (cl 1.1) means that where a correct construction of the contract documents leaves a matter regarding quality to the discretion of the architect, the contractor only fulfils its obligations if the architect is satisfied. It is suggested that any expression of dissatisfaction by the architect must be made within a reasonable period of the carrying out of the work and not left, eg, until the work has been built in. It should be noted that the Final Certificate is conclusive evidence that where the contract documents have expressly stated that the quality is to be to the approval of the architect, then the architect is so satisfied (cl 4.7.1). This would have the effect of preventing the employer bringing a claim regarding those items of work. The architect should therefore avoid using phrases such as 'to approval' or 'to the architect's satisfaction' in the contract documents, as it puts the employer at some risk.

3.10   If the phrase 'or otherwise approved' is used in a specification or Bill of Quantities this does not mean that the architect must be prepared to consider alternatives put forward by the contractor, nor that the architect must give any reasons for rejecting alternatives *(Leedsford v City of Bradford)*. It merely gives the architect the right to do so. A substitution would always constitute a variation whether or not this phrase is present in the specification.

Leedsford Ltd v The Lord Mayor, Alderman and Citizens of the City of Bradford (1956) 24 BLR 45 (CA)

In a contract for the provision of a new infant school the contract Bills stated 'Artificial Stone ... The following to be obtained from the Empire Stone Company Limited, 326 Deansgate, or other approved firm ...'. During the course of the contract the contractor obtained quotes from other companies and sent them to the architect for approval. The architect, however, insisted that Empire Stone was used and as Empire Stone was considerably more expensive the contractor brought a claim for damages for breach of contract. The court dismissed the claim stating 'The builder agrees to supply artificial stone. The stone has to be Empire Stone unless the parties agree some other stone, and no other stone can be substituted except by mutual agreement. The builder fulfils his contract if he provides Empire Stone, whether the Bradford Corporation want it or not; and the Corporation Architect can say that he will approve of no other stone except the Empire Stone' (Hodson LJ at page 58).

Obligations in respect of quality of sub-contracted work

3.11   Two methods of sub-contracting are allowed for under IFC98:
- sub-letting to a domestic sub-contractor selected by the main contractor, but with the written consent of the architect (cl 3.2);
- sub-letting to a named sub-contractor under the procedure set out in clause 3.3.1.

3.12   In both cases the contractor will still have ultimate responsibility for the standard of workmanship, materials, and goods provided by the sub-contractor (cl 3.3.9). With respect to named sub-contractors, the responsibility is qualified by clause 3.3.7. This makes it clear that the contractor is not responsible to the employer for design of the named sub-contract works, or the selection of materials and goods for those works, or the satisfaction of any performance specification or requirement for the sub-contract works, in so far as the named sub-contractor has or will be carrying out these tasks. This disclaimer applies whether or not the named sub-contractor is responsible to the employer. It is therefore extremely important that the named sub-contractor enters into a warranty with the employer, as otherwise the employer would have no redress against the sub-contractor should the design fail.

Compliance with statute

3.13   The contractor is under a statutory duty to comply with all legislation that is relevant to the carrying out of the work, eg in respect of goods and services, building and construction regulations, and health and safety. The duty is absolute and there is no possibility of contracting out of any of the resulting obligations.

3.14   IFC98 introduces a contractual duty in addition to the statutory duty. Clause 5.1 requires the contractor to 'comply with, and give all notices required by any statute, any statutory instrument, rule or order or any regulation or bylaw applicable to the Works". This obligation is limited by clause 5.3 which states that the contractor would not be contractually liable if the non-compliance resulted from carrying out work in accordance with the contract documents or further instructions issued by the architect. The contractor is reimbursed for any fees or charges, unless the tender documents made it clear that the contractor was to include for these in the contract sum (cl 5.1).

3.15   If the contractor finds any divergence between what the contract requires and statutory requirements, then the architect must be given immediate written notice (cl 5.2). The contractor is under no obligation to search for divergence, but is of course under a general obligation to use reasonable skill and care in carrying out the Works. The contract does not state what should happen once a divergence is discovered, but the architect should issue an instruction promptly to rectify the situation. The contractor may need to take immediate action in an emergency, but only in so far as is reasonably necessary to comply with statutory requirements (cl 5.4.1). Provided it was necessary, the work is treated as if it were a variation required by an instruction. There appears to be no obligation on the architect to cover the work with an instruction, but an accurate record should be kept.

Health and safety legislation

3.16   Clause 5.7.1 places a contractual obligation on the employer to ensure that the planning supervisor carries out his or her duties under the CDM Regulations. This is a wider obligation than the 'reasonable satisfaction with competence' obligation imposed by the CDM Regulations. Breach of it gives the contractor the right to determine the contract under clause 7.9.1(d). What is more likely to happen is that the contractor will claim for an extension of time or direct loss and/or expense for breach of clause 5.7.1, as this has been made a relevant event under clause 2.4.17 and a 'matter' under 4.12.9. An example might be where a planning supervisor delays in commenting on a contractor's proposed amendment to the Health and Safety Plan, and progress is thereby delayed.

3.17   Clause 5.7.2 also places a duty on the contractor, if acting as the principal contractor, to comply with all the relevant duties set out in the CDM Regulations. Breach of this duty is grounds for determination under clause 7.2.1(e). The warning notice still has to be given, and JCT Practice Note 27 suggests that the provision should only be used for situations where the Health and Safety Executive are likely to close the site. Any breach is covered, however, provided determination is not unreasonable or vexatious, and the architect and the employer should consider any breach that might lead to action being taken against the employer serious. If work needs to be postponed or other instructions given due to a breach by the contractor then there should be no entitlement to an extension of time or direct loss and/or expense.

3.18   The contractor is obliged to send any modification of the Health and Safety Plan to the employer (cl 5.7.2). This may occur, eg, due to unexpected site conditions or to changes in detailed design information from named sub-contractors. The contractor should take the cost of developing the Health and

Safety Plan into account at tender stage, and no claims could be made for adjusting it to suit the contractor or sub-contractor's working methods. If alterations are needed as a result of an instruction requiring a variation then the costs are included in valuing the variation, and the alterations may be taken into account in assessing an application for extension of time.

4.1   The date of possession or commencement and the date for completion of the work are key dates in any building contract, and IFC98 requires a 'Date of Possession' and a 'Date for Completion' to be inserted in the Appendix. It is good practice to give the contractor calendar dates at the time of tendering, and not vague indications such as 'to be agreed' or '8 weeks after approval by…'. The start date and the time of year that the work is carried will affect the contractor's costs, so that if only vague indications are given, it will impossible to compare tenders with accuracy. In addition, by the time it comes to executing the contract documents, the contractor may well protest that the dates now being proposed are not those that had been assumed in the tender, and may insist that the contract sum is adjusted.

4.2   In the event that work is started without proper agreement over dates, the contract will be subject to the Supply of Goods and Services Act 1982, which states that the work is completed within a reasonable time. It is unlikely, however, that once the job is underway the parties will be able to agree what a reasonable time might be.

4.3   As only one date for possession and completion is referred to in the printed form, if separate dates of possession and completion for different sections of the work are required, then the contract must be amended accordingly. The Sectional Completion Supplement sets out a schedule of amendments allowing for phased commencement and completion, and should always be used if this is required

4.4   The named sub-contractor must carry out its work in accordance with the programme agreed in NAM/T Section 1 clause 15 and Section II clause 1 (NAM/SC cl 12.1). It may be necessary for the sub-contractor to carry out its work in parts, in other words to return to site several times to suit the contractor's programme, in which case the details should be set out in NAM/T Section 1. This work may also have to be programmed across several sections if the Sectional Completion Supplement is executed.

Possession by the contractor

4.5   Possession of the site is a fundamental term of the contract. Failure to give contractor possession is a serious breach by the employer, which may amount to repudiation, and therefore give the contractor the right to treat the contract as at an end. Giving possession of only part of the site, or in stages, could amount to a breach unless this intention has been made clear in the contract documents *(Whittal Builders v Chester-le-Street DC)*. It is important to note that unless clause 2.2 has been stated to apply there is no relevant event allowing for an extension of time for failure to give full possession, therefore such failure may result in time being 'at large', ie the contractor will no longer be under an obligation to complete by the completion date

4.6   Degree of possession is such that there must be no interference that prevents the contractor from working in whatever way or sequence it chooses. With most jobs this means that the contractor must be given clear possession of the whole site up until Practical Completion. Where clear possession is not intended, then the tender documents should set out in detail the restrictions and the contract must be amended accordingly. Should the employer wish to use any part of the Works for any purpose during the time that the contractor has possession, this should also be made clear in the tender documents, otherwise it can only be with the agreement of the contractor (cl 2.1).

Whittal Builders Co Ltd v Chester-le-Street District Council (1987) 40 BLR 82
Whittal builders contracted with the council on JCT63 to carry out modernisation work to 90 dwellings. The contract documents did not mention the possibility of phasing but the council gave the contractor possession of the houses in a piecemeal manner. Even though work of this nature was frequently phased, the judge nevertheless found that the employers were in breach of contract for not giving the contractor possession of all 90 dwellings at the start of the contract, and the contractor was entitled to damages.

4.7   If clause 2.2 has been stated in the Appendix to apply, then it is possible for the employer to defer possession without the agreement of the contractor. The clause allows for deferment for a period not exceeding six weeks, and the maximum period required must be stated in the tender documents and inserted in the Appendix (cl 2.2). Any delay beyond the period stated in the Appendix is a breach of contract. It is the employer's right to defer possession, therefore the notice to the contractor should be written by the employer, on the advice of the architect. If possession is deferred, the contractor may claim an extension of time (cl 2.4.14) and direct loss and/or expense (cl 4.11(a)).

4.8   The parties are of course always free to renegotiate the terms of any contract. Therefore, if there is a delay in giving possession which is longer than the amount stated in the Appendix, the parties may have to agree a new date of possession, usually with a financial compensation to the contractor. Any further delay beyond the agreed date would of course be a breach.

4.9   Failure to give the contractor ingress to or egress of the site, in so far as this relates to land in the possession or control of the employer, is grounds for an extension of time (cl 2.4.12). The access need only be provided, however, to the extent that contract documents have indicated it will be available, or to the extent agreed between the architect and the contractor. Although not grounds for loss and expense or determination, this breach could give rise to a common law claim for damages, or even an allegation of repudiation if the access was fundamental to carrying out the work. It should be noted that this clause cannot be used to grant an extension of time where access is not available to part of the site itself – this can only be dealt with under clause 2.2.

Progress

4.10 It would normally be implied into a construction contract that a contractor will proceed 'regularly and diligently', and this is an express term in IFC98 (cl 2.1). The contractor is free to organise its own working methods and sequences of operations, with the qualification that it must comply with statutory requirements and the Health and Safety Plan. This has been held to be the case even where the contractor's chosen sequencing may cause extra cost to the employer with the operation of fluctuation provisions *(GLC v Cleveland Bridge and Engineering)*.

Greater London Council v Cleveland Bridge and Engineering Co (1986) 34 BLR 50 (CA)
The GLC employed Cleveland Bridge to fabricate and install gates and gate arms for the Thames Barrier. The specially drafted contract provided dates by which Cleveland Bridge had to complete certain parts of the works. Clause 51 was a fluctuations provision which allowed for adjustments to be made to the contract sum if, for example, the rates of wages or process of material rose or fell during the course of the contract. The clause also stated 'provided that no account shall be taken of any amount by which any cost incurred by the Contractor has been increased by the default or negligence of the contractor'. The contract was lengthy, and Cleveland Bridge left a part of the work to be done at the very end of the period, but delivered the gates on time. The result was that the GLC had to pay a large amount of fluctuations in respect of the work done at the last minute. The GLC argued that the contractor had failed to proceed regularly and diligently, and therefore was in default. The court held that even if the slowness of the contractor's progress might at certain points have given the employer the right to determine the contract under the determination provisions, this would not by itself be a breach of contract as referred to in clause 51. The contractor could organise the work any way it wished provided it completed on time, and was therefore owed the full amount of the fluctuations.

4.11   There is no requirement under IFC98 for the contractor to produce a master programme. Of course there is nothing to prevent such a requirement being introduced through the Bill of Quantities or specification, but it should be made clear that this will not be a contract document. If it is to be required, then it is wise to require that the contractor submits it before the contract is entered. It is notoriously difficult to extract programmes from contractors once work has commenced – sometimes the programme when finally produced can include an element of post-rationalisation to show that early events caused problems and delays! If requested, it is also advisable to ask for a critical path analysis as otherwise assessing the effect of delays can be quite difficult.

4.12   One of the functions of a programme is to indicate when information will be needed from the architect, and if the information release schedule is not used it would be open to the contractor to submit such a programme, even if one were not requested. Again, the programme would not be contractually binding. It should be remembered that even if the contractor's programme shows an intention to complete early, there is no implied duty on the employer to enable the contractor to achieve this early completion *(Glenlion Construction v The Guinness Trust)*, and in particular the architect's obligation to provide information would not be assessed against this programme. It would be sensible, however, to notify the employer if an early finish is shown, as the employer should be alert to the possibility that he or she may receive the building earlier than had been anticipated.

Glenlion Construction Ltd v The Guinness Trust (1987) 39 BLR 89
The Trust employed Glenlion Construction to carry out works in relation to a residential development at Bromley, Kent. The contract was on JCT63, which required the contractor to complete 'on or before' the date for completion, and to provide a programme. Disputes arose which went to arbitration and several questions of law regarding the contractor's programme were subsequently raised in court. The contractor later claimed loss and expense on the ground that it was prevented from economic working and achieving the early completion date shown on its programme only by failure of the architect to provide necessary information and instructions to the dates shown. The court decided that Glenlion was entitled to complete before the date for completion whether or not it was contractually bound to produce a programme and whether or not it did in fact produce one. Glenlion was therefore entitled to carry out the works in a way which would achieve an earlier completion date. However there was no implied obligation that the employer (or the architect) should perform its obligations so as to enable the contractor to complete by any earlier completion date shown on the programme.

Completion

4.13 The most important reason for giving an exact completion date in a building contract is that it provides a fixed point from which damages may be payable in the event of non-completion. Generally in construction contracts the damages are 'liquidated', and typically are fixed at a rate per week of overrun.

4.14 The contractor is obliged to complete the Works by the completion date, and in general accepts the risk of all events that might prevent completion by this date, except to the extent that the contract provides otherwise. The contractor would normally be relieved of this obligation if the employer caused delays or in some way prevented completion. In order to avoid situations arising where the contractor is no longer bound by the completion date, most contracts contain provisions allowing for the adjustment of the completion date in the event of certain delays caused by the employer.

4.15 In contracts it is sometimes essential that completion is achieved by a particular date and failure would mean that the result is worthless. This is sometimes referred to as 'time is of the essence'. Breach of such a term would be considered a fundamental breach, and would give the employer the right to terminate performance of the contract, and treat all its own obligations as at an end. The expression 'time is of the essence' is seldom, if ever, applicable to building contracts such as IFC98, as the inclusion of extensions of time and liquidated damages provisions imply that the parties intended otherwise.

4.16 In IFC98 a 'Date for Completion' is inserted in the Appendix, which is the date agreed at the time of entering into the contract (if the Sectional Completion Supplement is used a separate date will be stated for each section). IFC98 provides for the granting of extensions of time. This does not take the form of fixing a new 'Date for Completion', which remains as originally agreed, but is referred to as extending time for the completion of the Works. There is no provision for reducing the contract period to a date earlier than the date for completion, even when substantial work is omitted (cl 2.3). If the contractor fails to complete by the Date for Completion, or any extended time fixed under clause 2.3, liquidated damages become payable (see figure 4).

Figure 4 Completion and liquidated damages

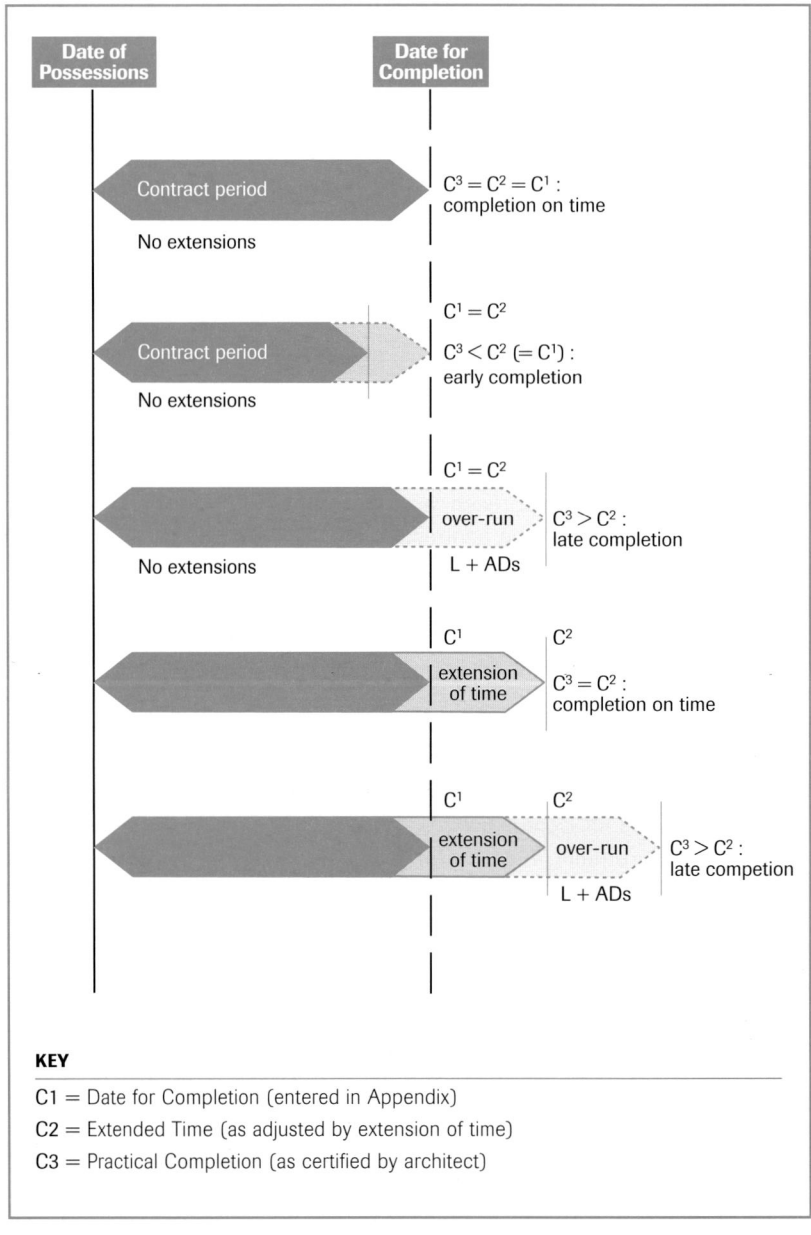

KEY

C1 = Date for Completion (entered in Appendix)

C2 = Extended Time (as adjusted by extension of time)

C3 = Practical Completion (as certified by architect)

Extensions of time

Principle

4.17   The main reason for including extension of time provisions in a building contract is to preserve the employer's right to liquidated damages, in the event that the contractor fails to complete on time due in part to some action for which the employer is responsible. If no such provisions were included, and a delay occurred that was caused by the employer, this would in effect be a breach of contract by the employer and the contractor would no longer be bound to complete by the completion date *(Peak v McKinney)*. The employer would therefore lose the right to liquidated damages, even though much of the blame for the delay rests with the contractor. The phrase 'time at large' is often used to describe this situation, however this is strictly speaking a misuse of the phrase as in most cases the contractor would remain under an obligation to complete within a reasonable time.

Peak Construction (Liverpool) Ltd v McKinney Foundations Ltd (1970) 1 BLR 111 (CA)

Peak Construction were main contractors on a contract to construct a multi-storey block of flats for Liverpool Corporation. The main contract was not on any of the standard forms, but a contract drawn up by the council. McKinney Foundations were nominated sub-contractors to design and construct the piling. After the piling was complete and the sub-contractors had left the site, serious defects were discovered in one of the piles, and following further investigation minor defects were found in several other piles. Work was halted while the best strategy for remedial work was debated between the parties. The city surveyor did not accept the initial remedial proposals, and it was agreed that an independent engineer would prepare a proposal. The council refused to agree in advance to accept his decision, and delayed making the appointment. Altogether it was 58 weeks before work resumed, although the remedial work took only six weeks, and the main contractors brought a claim against the sub-contractors for damages. The Official Referee at first instance found that the entire 58 weeks delay was caused by the nominated sub-contractor and awarded £40,000 damages for breach of contract, based in part on liquidated damages, which the council had claimed from the contractor. McKinney appealed, and the Court of Appeal found that the 58 weeks could not possibly be all due to the breach of the sub-contractor, but was in part caused by the tardiness of the council. This being the case, and as there were no provisions in the contract for extending time for their delay, the council lost their right to claim liquidated damages, and this component of the damages awarded against the sub-contractor was disallowed. Even if the contract had contained such a provision, the failure of the architect to exercise it would have prevented the council from claiming liquidated damages. The only remedy would have been for the council to prove what damages they had suffered as a result of the breach.

Procedure

4.18   In IFC98 the provisions for granting an extension of time are under clause 2.3. In the Sectional Completion Supplement the provisions are modified to apply independently to each section. The contractor must give written notice 'forthwith' to the architect, when it appears that progress is being or is likely to be delayed. The notice must be given whether or not completion is likely to be delayed whatever the cause might be (ie the requirement to give a notice is not limited to circumstances where the contractor is claiming an extension of time). The notice should set out the cause of the delay, but the contractor is not required to identify whether the cause is an event listed in clause 2.4, and the architect's obligation to issue an appropriate extension of time is not dependent on the contractor having done so. However, the contractor is required to provide the architect with any information reasonably required.

4.19   Following notification, the architect must then assess the delay caused and issue an extension of time if appropriate. The extension can only be given in relation to delay caused by events listed in clause 2.4 (Relevant Events), and the application of some of these is conditional upon it having been stated to apply in the Appendix. Although the point has not been decided by the courts in relation to IFC98, it appears that prior to Practical Completion notification is a condition precedent to the award of an extension of time, in other words the architect may not issue an extension unless a valid notice has been given. In any event, it would be difficult to make an assessment in the absence of any information from the contractor, and the architect would be unwise to make such an extension without taking expert advice.

4.20   In regard to Relevant Events the following points should be noted:
- 'force majeure' (cl 2.4.1) is a French term used 'with reference to all circumstances independent of the will of man, and which it is not in his power to control'. It includes Acts of God and other matters outside the control of the parties. However, many items under this category, eg strikes, fire and weather, are dealt with elsewhere in the contract;
- weather is to be exceptional and adverse (ie not that which would be expected at the time of year in question) (cl 2.4.2). The effect of the weather is assessed at the time the work is actually carried out, not when it should have been, according to the contractor's programme;
- Specified Perils (cl 2.4.3) can under certain circumstances include events caused by the contractor's own negligence;
- the very wide protection afforded with respect to strikes, and not simply those directly affecting the Works, but also those causing difficulties in preparation and transportation of goods and materials. Such strikes will not

necessarily be confined to the UK and given the current extent of overseas imports, the effects could be considerable (**cl 2.4.4**). It was generally considered, however, that the recent fuel tax blockades was not an event which fell under this clause (or indeed any other 'Relevant Event');

- failure to supply information now relates to both the information release schedule, and to such information as is necessary 'having regard to the Progress of the Works', whether or not specifically applied for in writing (**cl 2.4.7.2**);
- shortage of labour and materials (**cl 2.4.10 and 2.4.11**) is only a Relevant Event if stated in the Appendix to apply – and only for things beyond the contractor's control and not reasonably forseeable at the base date;
- it is generally considered that 'failure to give ingress to or egress from' the site (**cl 2.4.12**) does not cover the situation where the contractor is not given possession of part of the site;
- local authority or statutory undertaker's work (**cl 2.4.13**) – this only covers situations where work is done in pursuance of statutory duties. If directly engaged by the employer this comes under clause 2.4.8.

4.21   There is no time limit on when the decision should be made. However the clause states 'so soon as he is able', and as failure to grant an extension properly due could result in time being 'at large', the architect should take care to deal with the matter reasonably quickly. It certainly would be unwise to set the matter to one side until the end of the project. The clause does not require the architect to notify the contractor if the decision is that no extension of time is due, but it would be reasonable to make the contractor aware of the position. There is also no obligation to explain why any extension has been awarded.

4.22   The architect may award further extensions of time in respect of certain events which occur after the Date for Completion or any extended date, ie when the contractor is in 'culpable delay' *(Balfour Beatty v Chestermont Properties)*. In this event the extension is added on to the date that has passed, referred to as the 'net' method of extension. The range of events is less than those for which an extension can be awarded before the date for completion has passed: it does not include many neutral events such as delays by the Specified Perils, or the contractor's inability for reasons beyond his control to secure necessary labour.

Balfour Beatty Building Ltd v Chestermont Properties Ltd (1993) 62 BLR 1
In a contract on JCT80 the works were not completed by the revised completion date and the architect issued a non-completion certificate. The architect then issued a series of variation instructions and a further extension of time which had the effect of fixing a

completion date two and a half months before the first of the variation instructions. He then issued a further non-completion certificate and the employer proceeded to deduct liquidated damages. The contractor took the matter to arbitration and then appealed certain decisions on preliminary questions given by the arbitrator. The court held that the architect's power to grant an extension of time pursuant to clause 25.3.1.1 could only operate in respect of relevant events that occurred before the original or the previously fixed completion date, but the power to grant an extension under 25.3.3 applied to any relevant event. The architect was right to add the extension of time retrospectively (termed the 'net' method).

4.23   The architect may make a further extension of time at any time up to 12 weeks after Practical Completion (cl 2.3), and may review any extension of time previously given The final review may extend the date previously fixed but may not bring it forward, although the architect may be able to offset a reduction or omission of work against the effect of another delaying event. It is clear that at this point there is no requirement for the notification by the contractor. There is no requirement to notify the contractor if it is decided that no further extension is due, but again it would be reasonable to do this.

Assessment

4.24   The contractor is entitled to extensions of time properly due under the contract and any failure on the part of the architect to administer the provisions correctly would constitute a breach of contract on the part of the employer. On the other hand, the architect has no power to grant extensions of time except as provided for in the contract, and for delays caused by events as listed in clause 2.4.

4.25   The architect should make an objective assessment of every notice received. The aim is to establish if a delay has been caused by the event cited, whether the delay is likely to disrupt the programme and consequently delay the final completion date, and if so assess the probable extent of that final delay. The contractor's programme can be used as a guide and may be particularly useful where the programme shows a critical path, but although it may be persuasive evidence it is not conclusive or binding. The effect on progress is assessed in relation to the work being carried out at the time of the delaying event, rather than the work that was programmed to be carried out.

4.26   Clause 2.3 contains the important proviso that the contractor must use its 'best endeavours' to prevent delay. The proviso refers to preventing delay caused by a relevant event, not to preventing the event itself. The architect can assume, therefore, that the contractor will take steps to minimise the effect of the delay on the completion date, eg through re-programming the remaining works. The phrase 'best endeavours' appears to suggest something more than 'reasonable' or 'practicable' but it is unlikely to extend to excessive expenditure. It also states that the contractor 'shall do all that may be reasonably required to the satisfaction of the Architect/ the Contract Administrator to proceed with the Works'. However if the architect requires measures which amount to a variation, then this may result in a claim for loss and/or expense.

4.27 The effects of any delay on completion – taking into account the contractor's 'best endeavours' (cl 25.3.4) – are not always easy to predict. The architect is required to reach an opinion, and in doing this the architect owes a duty to both parties to be fair and reasonable This applies even where the delay has been caused by the architect, for example where the architect has failed to issue drawings within the time limits stipulated in the contract.

4.28 It sometimes happens that two or more delaying events can happen simultaneously, or with some overlap, and this can raise difficult questions with respect to the awarding of extensions of time. In the case of concurrent delays involving two or more clause 2.4 events, it has been customary to grant the extension in respect of the dominant reason, but this is only appropriate where the dominant reason begins before, and ends after, any other reasons. Even then, if the dominant reason is not a ground for loss and expense, this may still be due in respect of the other delaying events *(H Fairweather & Co v Wandsworth)*.

4.29 Where one overlapping delaying event is a clause 2.4 event and the other is not, in other words one is the employer's risk and the other the contractor's, a difficult question arises as to what extension of time is due. It would seem logical that the contractor should be given an extension of time for the full length of delay caused by the clause 2.4 event, irrespective of the fact that during the overlap the contractor was also causing delay. Taking any other approach, by eg splitting the overlap period and awarding only half to the contractor, could result in the contractor being subject to liquidated damages for delay partly caused by the employer.

H Fairweather & Co Ltd v London Borough of Wandsworth (1987) 39 BLR 106
Fairweather entered into a contract with the London Borough of Wandsworth to erect 478 dwellings. The contract was on JCT63. Pipe Conduits Ltd were nominated sub-contractors for underground heating works. Disputes arose and an arbitrator was appointed who made an interim award. Leave to appeal was given on several questions of law arising out of the award. The arbitrator had found that where a delay occurred

which could be ascribed to more than one event the extension should be granted for the dominant reason. The dominant reason was strike action, and the arbitrator had granted an extension of 81 weeks for this reason, and made it clear that this reason did not carry any right to claim for direct loss and/or expense. The court stated that an extension of time was not a condition precedent for an award of direct loss and/or expense, and that the contractor would be entitled to direct loss and/or expense for other events which had contributed to the delay.

Partial possession

4.30   The employer may take possession of completed parts of the Works ahead of Practical Completion by operating clause 2.11. 'Partial possession' requires the agreement of the contractor, which cannot be unreasonably withheld, although as the contractor will not be able to claim later an extension of time or losses due to any disruption, it would not be unreasonable for the contractor to withhold consent unless certain that the partial possession will not affect operations on the rest of the site. Practical Completion is 'deemed to have occurred' for the 'relevant part' of the Works, the defects liability period for that part is deemed to have commenced, and the certificate of making good defects has to be issued for that part separately However, it would appear that that this remains part of the Works, and is still to be included under the certificate of Practical Completion. It is notable that the clause does not state that the Works to that part must have reached Practical Completion, but in view of the contractual consequences it would be unwise for the employer to take possession before they have *(see para 4.38 below)*.

4.31   Liquidated damages are reduced by the proportion of the value of the 'relevant part' of the Works to the contract sum and half of the retention is released for that proportion of the Works. If clause 6.3C applies the employer is responsible for insuring the possessed part under clause 6.3C.1. If clause 6.3A or 6.3B applies the employer may wish to consider insuring the part as the contractor's obligation to insure the Works will cease .

4.32   To bring the provision into operation the architect must issue a written statement to the contractor identifying precisely the extent of the 'relevant part' and the date of possession (the 'Relevant Date'). This should be done with great care, even using a drawing to illustrate the extent, and communicating to the insurers where relevant. The statement must be issued immediately after the part is taken into possession, but in practice it would be wise to circulate the drawings and information in advance, so that all parties are clear as to the details of what will occur.

Use or occupation before Practical Completion

4.33 Clause 2.1 provides for the situation where the employer wishes to 'use or occupy the site or the Works or any part' when the contractor is still in possession. The purposes for which the employer might require this are described simply as 'storage of his goods or otherwise', and while this might suggest a limited role was intended, in theory at least the clause places no limits on what form it might take. In practice, as the consent of the contractor in writing is required, and like partial possession discussed above, it appears not to be included as an event for which an extension of time may be awarded, or as a matter for which losses for disturbance can be claimed, the contractor would not be unreasonable in withholding permission unless certain the intervention would cause no delays. Before the contractor is required to give consent the employer or the contractor as appropriate must notify the insurers. If any additional premium is required this is added to the contract sum (cl 2).

4.34 There can be situations where the contractor has not completed by the Date for Completion and, although no sections of the Works are sufficiently complete to allow the employer to take possession of those parts under clause 18, the employer is nevertheless anxious to occupy at least part of the Works. There is nothing in the contract that allows for this. A suggestion was put forward in the 'Practice' section of the *RIBA Journal* (February 1992) which is frequently found useful in practice (see figure 5). In this arrangement, in return for being allowed to occupy the premises, the employer agrees not to claim liquidated damages during the period of occupation. Practical Completion obviously cannot be certified, and there is no release of retention money until it is. Matters of insuring the Works will need to be settled with the insurers. Because such an arrangement would be outside the terms of the contract it should be covered by a properly drafted agreement which is signed by both parties. It may also be sensible to agree that in the event that the contractor still fails to achieve Practical Completion by the end of an agreed period, liquidated damages would run again, possibly at a reduced rate. In most circumstances this arrangement would be of benefit to both parties, and is far preferable to issuing a heavily qualified certificate of Practical Completion listing 'except for' items.

Practical Completion

4.35 The architect is obliged to certify Practical Completion (cl 2.8) when in the architect's opinion the following two criteria are fulfilled:
- Practical Completion of the Works is achieved *(see below)*;
- the contractor has sufficiently complied with clause 5.7.4 (supply of information required for the Health and Safety File).

Figure 5 Practice Section, RIBA Journal *(Feb, 1992)*

CONTRACTS

Employer's possession before Practical Completion under JCT contracts

It is not uncommon for the Employer after the completion date has passed to wish to take possession of the Works before the Contractor has achieved Practical Completion. In this event an ad hoc agreement between Employer and Contractor is required to deal with the situation.

In respect of such an agreement members may wish to have regard to the following note . . .

Outstanding items

Where it is known to the architect that there are outstanding items, practical completion should not be certified without specially agreed arrangements between the employer and the contractor. For example, in the case of a contract where the contract completion date has passed it could be so agreed that the incomplete building will be taken over for occupation, subject to postponing the release of retention and the beginning of the defects liability period until the outstanding items referred to in a list to be prepared by the architect have been completed, but relieving the contractor from liability for liquidated damages for delay as from the date of occupation, and making any necessary changes in the insurance arrangements.

In such circumstances either the Certificate of Practical Completion form should not be used or it should be altered to state or refer to the specially agreed arrangements. In making such arrangements the architect should have the authority of the client-employer.

When the Employer is pressing for premature practical completion there is a need to be particularly careful where there are others who are entitled to rely on the issue of a Practical Completion Certificate and its consequences. In the case where part only of the Works is ready for hand-over the partial possession provisions* can be operated to enable the Employer with the consent of the Contractor to take possession of the completed part.

*JCT 80: clause 18
IFC 84: clause 2.11 in appendix to JCT Practice Note IN/1 (if incorporated in the contract)

4-36 Clause 12.8 then continues by stating that 'Practical Completion of the Works shall be deemed for all the purposes of the contract to have taken place on the day named in such certificate'. Although the wording of clause 2.8 is somewhat circular, in that it is in effect saying that 'Practical Completion of the Works' is to be certified when 'Practical Completion of the Works' has taken place plus another event, it is suggested that a correct analysis of this clause is that Practical Completion only occurs when both conditions are met, the principal argument for this being that only one date is entered on the certificate. The architect would be entitled to withhold the certificate until all significant health and safety information has been received, even if the actual Works have been finished for some time, and would certainly be entitled to do so if the lack of information put the employer at risk of being in breach of the CDM Regulations. It should be noted that the contractor's obligation to supply information for the Health and Safety File depends on the planning supervisor having requested it in writing. Even then the use of the term 'complied sufficiently' may allow the architect to use his or her discretion in issuing the certificate with some information missing.

4.37 Deciding when the Works have reached Practical Completion often causes some difficulty. It is suggested that Practical Completion means the completion of all works required under the contract and by subsequent instruction. Although it has been held that the architect has a discretion to certify Practical Completion where there are very minor items of work left incomplete, on 'de minimus' principles *(H W Neville (Sunblest) v William Press)*, this discretion should be exercised with extreme caution. Contrary to the opinion of many contractors, there is no obligation to issue the certificate when the project is 'substantially' complete, or even when it is capable of occupation by the client, if there are items still outstanding.

H W Neville (Sunblest) Ltd v William Press & Son Ltd (1981) 20 BLR 78
William Press entered into a contract with Sunblest to carry out foundations, groundworks and drainage for a new bakery on a JCT63 contract. A practical completion certificate was issued, and new contractors commenced a separate contract to construct the bakery. A certificate of making good defects and a final certificate were then issued for the first contract, following which it was discovered that the drains and the hard standing were defective. William Press returned to site and remedied the defects, but the second contract was delayed by four weeks and Sunblest suffered damages as a result. It commenced proceedings, claiming that William Press was in breach of contract and in their defence William Press argued that the plaintiffs were precluded from bringing the claim by the conclusive effect of the final certificate. Judge Newey decided that the final

certificate did not act as a bar to claims for consequential loss. In reaching this decision he considered the meaning and effect of the certificate of practical completion and stated 'I think that the word "practically" in clause 15(1) gave the architect a discretion to certify that William Press had fulfilled its obligation under clause 21(1) where very minor de-minimus work had not been carried out, but that if there were any patent defects in what William Press had done then the architect could not have issued a certificate of practical completion' (at page 87).

4.38   The reason for proceeding with extreme caution is the considerable complications that can arise as a result of premature certification. Even though the employer, anxious to move into the newly completed works, may be pressing for early completion, and the contractor, anxious to avoid liquidated damages, may be even more enthusiastic, the temptation to issue the certificate, particularly one qualified by long schedules of outstanding work, should be resisted. The architect should explain to the employer that they would be in a difficult position contractually, as the following contractual problems will remain unresolved:

- half of the retention will be released, leaving only half in hand (cl 4.3). This puts the employer at considerable risk, as the 2.5 per cent is only intended to cover latent defects;
- the defects liability period begins (cl 2.10);
- the onus shifts to the architect to notify the contractor of all necessary outstanding work under clause 2.10. If the architect fails to include something the contractor would have no authority to enter the site to carry it out – therefore the architect will inevitably become involved in managing and programming the outstanding work;
- the site is now in the possession of the employer, and (unless clause 6.3C or 6.3B applies) the contractor will no longer cover the insurance of the Works. The insurers will need to be informed about the programme for the outstanding work;
- the contractor's liability for frost damage ends (cl 2.10);
- the contractor's liability for liquidated damages ends;
- regular interim certificates cease to be issued (cl 4.2);
- the employer will be the 'occupier' for the purposes of the Occupiers' Liability Act 1957 and may also be subject to claims regarding health and safety.

4.39   The certificate must be issued as soon as the criteria in clause 2.9 are met. The contractor is obliged to complete 'on or before' the completion date and once Practical Completion is certified the employer is obliged to accept the

Works. Employers who wish to accept the Works only on the date in the contract would need to amend the wording. If the Sectional Completion Supplement is executed, Practical Completion must be certified for each section of the Works.

Procedure at Practical Completion

4.40   The contract sets out no procedure for what happens at Practical Completion, it simply requires the architect to certify it. The contract Bills may set out a procedure, and the architect should check carefully at tender stage to ensure that the procedure is satisfactory.

4.41   Leading up to Practical Completion it appears to be widespread practice for architects to issue 'snagging' lists, sometimes in great detail and on a room-by-room basis. The contract does not require this, and neither do most standard terms of appointment. Under the contract, responsibility for quality control and snagging rests entirely with the contractor. In adopting this role the architect may be assisting the contractor, and although this may appear to benefit the employer it may lead to confusion over the liability position, which could cause problems at a future date. If the architect feels that the Works are not complete there is no obligation to justify this opinion with schedules of outstanding items. It is suggested that the best course may be to draw attention to typical items, but to make it clear that the list is indicative and not comprehensive.

4.42   It is frequently the practice for the contractor to arrange a 'handover' meeting. The term is not used in IFC98 and although handover meetings can be of use, particularly in introducing the finished project to the employer, the fact that one has been arranged or taken place is of no contractual significance. Even where a meeting has taken place at which the employer has expressed approval of the Works, or the contractor has stated in writing that the Works are complete, it remains the architect's responsibility to decide when Practical Completion has been achieved.

Failure to complete by the Date for Completion

4.43   In the event of failure to complete by the Date for Completion or any extended date, the architect is required to certify this fact (cl 2.6). It should be noted that the issue of the certificate is an obligation on the architect and not a matter of discretion. It should be issued promptly, as the certificate is a condition precedent to deduction of liquidated damages (cl 2.7). If the Sectional Completion Supplement is used, a separate certificate will be needed for each incomplete section. Once the certificate has been issued the contractor is said to be in 'culpable delay'. The employer, provided that it has issued the necessary notices, may then deduct the damages from the next interim certificate, or reclaim the sum as a debt. Note that fluctuations provisions are frozen from this point. If a new completion date is later set this has the effect of canceling the non-completion certificate and the architect must issue a further non-completion certificate if necessary.

Liquidated and ascertained damages

4.44   The agreed rate for liquidated and ascertained damages is entered in the Appendix. This is normally expressed as a specific sum per week (or other unit) of delay, to be allowed by the contractor in the event of failure to complete by the Date for Completion (note there may be several different rates where the Works are divided into sections). The amount must be calculated on the basis of a genuine pre-estimate of the loss likely to be suffered. Provided that it is, the sum will be recoverable without the need to prove the actual loss suffered, and irrespective of whether the actual loss is significantly less or more than the recoverable sum. In other words, once the rate has been agreed, both parties are bound by it. If 'nil' is inserted then this may preclude the employer from claiming any damages at all *(Temloc v Erril)*, whereas if the Appendix is left blank the employer may be able to claim general damages.

Temloc Ltd v Erril Properties (1987) 39 BLR 30 (CA)

Temloc entered into a contract with Erril Properties to construct a development near Plymouth. The contract was on JCT80 and was in the value of £840,000. '£ nil' was entered in the Appendix against clause 24.2, liquidated and ascertained damages. Practical completion was certified around six weeks later than the revised date for completion. Temloc brought a claim against Erril Properties for non-payment of some certified amounts, and Erril counter-claimed damages for late completion. It was held by the court that the effect of '£ nil' was not that the clause should be disregarded (because, for example, it indicated that it had not been possible to assess a rate in advance), but that it had been agreed that there should be no damages for late completion. Clause 24 is an exhaustive remedy and covers all losses normally attributable to a failure to complete on time. The defendant could not, therefore, fall back on the common law remedy of general damages for breach of contract.

4.45   Before liquidated and ascertained damages may be claimed the following preconditions must have been met:
- the contractor must have failed to complete the works by the Date for Completion or any extended date;
- the architect must have issued a non-completion certificate;
- the architect must have fulfilled all duties with respect to the award of an

extension of time;
- the employer must have given the contractor written notice of its intention.

4.46   The written notice to be given depends on whether the employer intend to deduct liquidated damages from monies due, or to require the contractor to pay the sum to the employer. If the employer wishes to deduct liquidated damages from an amount payable on a certificate, clause 2.7.2 states that the employer must give a notice pursuant to clause 4.2.3(b) or 4.6.1.3 (this provision derives from the requirements of section 111 of the Housing Grants, Construction and Regeneration Act 1996). If the employer wishes to recover the amount as a debt then this has to be 'required in writing' and the notification must be given to the contractor no later than five days before the final date for payment of the debt due under the Final Certificate (cl 2.7.1).

4.47   It is notable that IFC98 refers in both clause 2.7.1 and 2.7.2 to the liquidated damages recoverable for the period 'during which the Works shall remain or have remained incomplete', and not for the period up until the certified date of Practical Completion of the Works. There is at least room for argument that once the Works are complete, even if the certificate has been withheld due to the lack of provision of health and safety information, the liquidated damages should cease. It is likely that the ambiguity is due to historical accident, as the CDM Regulations provisions came later, but it is a pity that the JCT have not adjusted the liquidated damages provisions accordingly. It is suggested that the employer would be entitled to levy the damages while the lack of information was putting the employer in possible breach of the CDM Regulations should it occupy the building. The employer, after all, would be suffering exactly the same losses as would be suffered due to the Works being incomplete. As this must have been apparent to both parties at the time the contract was entered into, it would be reasonable to proceed on that basis.

4.48   If an extension of time is given following the issue of a certificate of non-completion then this has the effect of cancelling that certificate. A new non-completion certificate must be issued if the contractor then fails to complete by the new completion date (cl 2.6). The employer must, if necessary, repay any liquidated damages recovered for the period up to the new completion date (cl 2.8). Clause 2.7 also states that any notice issued under that clause by the employer shall 'remain effective unless withdrawn by the Employer', notwithstanding that a further extension of time has been granted. It is suggested that nevertheless if the intention is to withhold payments due and the deduction was postponed to a later certificate then fresh notices should be issued. In *Department of Environment for Northern Ireland v Farrans* it was decided that the contractor has the right to interest on any repaid liquidated damages. This decision, however, concerned JCT63 and as IFC98 clause 2.8 expressly refers to repayment without stipulating that interest is due, it is generally considered that interest is not due in this situation.

Department of Environment for Northern Ireland v Farrans (Construction) Ltd (1981) 19 BLR 1 (NI)

Farrans was employed to build an office block under JCT63. The original date for completion was 24 May 1975, but this was subsequently extended to 3 November 1977. During the course of the contract the architect issued four certificates of non-completion. By 18 July 1977 the employer had deducted £197,000 in liquidated damages, but following the second non-completion certificate repaid £77,900 of those deductions. This process was repeated following the issue of the subsequent non-completion certificates. Farrans brought proceedings in the High Court of Justice in Northern Ireland, claiming interest on the sums that had been subsequently repaid. The court found for the contractor, stating that the employer had been in breach of contract in deducting monies on the basis of the first, second and third certificates, and that the contractor was entitled to interest as a result. The BLR commentary should be noted, which questions whether a deduction of liquidated damages empowered by clause 24.2 can retrospectively be considered a breach of contract, but the case has not been overruled.

4.49   Certificates should always show the full amount due to the contractor. It is the employer alone that makes the deduction of liquidated damages. The employer would not be considered to have waived its claim by a failure to deduct damages from the first or any certificate under which this could validly be done, and would always be able to reclaim them as a debt at any point up until the final certificate.

5.1   The architect's authority to administrate the contract derives from its wording. The contract places various duties on the architect, eg to supply necessary information, and to issue certificates or statements, and it also confers on the architect a wide variety of powers, eg the power to issue instructions (see figures 6 and 7). In some matters the architect will act as agent of the employer, eg when issuing instructions which vary the Works, and in others as an independent decision-maker, eg when deciding on claims for direct loss and/or expense. Where making a decision between the parties it would be implied that the architect must act fairly at all times.

Figure 6 Key powers of the architect

JCT98	Key powers of the architect
1.10	Consent to removal of goods from site
2.10	Instruct defects are not to be made good
3.2	Consent to domestic sub-contractors
3.3.2	Issue instructions naming a person as sub-contractor
3.6	Issue instructions requiring a variation
3.6	Sanction a variation by the contractor
3.9	Instruct errors in setting out can remain
3.12	Require contractor to open up work/ get work tested
3.14.1	Issue instructions requiring removal of work not in accordance with contract documents
3.14.2	Issue instructions regarding work not carried out in a proper and workmanlike manner
3.15	Issue instructions postponing work
6.2.4	Instruct 6.2.4 insurance is taken out
6.3D.1	Instruct 6.3D insurance is taken out
7.2.1	Give contractor notice of default(s)

Figure 7 Key duties of the architect

JCT98	Key duties of the architect
1.4	Issue instructions regarding discrepancies/ divergencies of contract
1.6	Provide contractor with copies of contract documents
1.6	Provide contractor with further copies of contract drawings and specification/schedules of work/contract bills
1.7.1	Provide contractor with copies of information referred to in the Information Release Schedule
1.7.2	Provide contractor with such further drawings as are reasonably necessary and issue necessary instructions
1.9	Send duplicate copies of certificates to contractor
2.3	Give an extension of time (if contractor submits a notice and architect considers completion date will be delayed, and delay caused by a Relevant Event)
2.6	Issue certificate(s) of non-completion
2.9	Certify Practical Completion
2.10	Issue Schedule of Defects
2.10	Issue Certificate of Making Good Defects
2.11	Issue statement regarding Partial Possession (on employer's behalf)
3.3.1-3.3.3	Issue instructions with regard to named sub-contractors
3.5.1	Issue instructions in writing
3.5.2	Comply with contractor's request for empowering provision
3.8	Issue instructions regarding Provisional Sums
3.9	Determine any levels that may be required
4.2	Certify interim payments
4.3	Certify payment on Practical Completion
4.6.1.1	Issue Final Certificate
4.11	Ascertain or instruct QS to ascertain any amounts of direct loss and/or expense incurred
5.2	Issue instructions regarding any discrepancies/ divergencies discovered between contract documents and Statutory requirements
6.2.4	Where required, instruct contractor to take out joint names policy
6.3D.1	Where required, instruct contractor to obtain quotation and take out insurance

5.2 Failure by the architect to comply with any obligation, either express or implied, may result in the contractor suffering losses. The architect is not a party to the contract, therefore if the contractor wishes to bring a claim, this would in the first instance have to be against the employer. The architect's obligations to the employer derive, of course, from the terms (either express or implied) of the professional appointment as agreed with the employer. It is likely, however, that any failure to administer the building contract according to its terms would be a breach of the architect's duties to the employer, and therefore the employer may seek in turn to recover the losses from the architect.

5.3 Direct control over the carrying out of the contract Works, including the manner in which the Works are undertaken, is solely the responsibility of the main contractor under the contract. The architect will normally inspect the work at intervals. The duty to inspect arises not from IFC98, which includes no express provision relating to inspection or monitoring of work by the architect, but directly from the terms of appointment. Clearly when the architect is required under the contract to form an opinion on various matters, including the standard of work and materials prior to issuing a certificate, then it would be implied, even if not expressly set out in the terms of appointment, that some form of inspection must take place.

Person in charge

5.4 The contractor is required to keep on the site 'at all reasonable times' a competent 'person-in-charge' (cl 3.4). What would be reasonable would depend on the nature and scale of the project, but as the JCT has chosen not to use the term 'constantly' as in JCT98, it would appear that something less than full time presence would be acceptable. There is no requirement in the contract Conditions to have the person named, but as this person may receive any instructions given by the architect, and therefore acts as the contractor's agent, it would be good practice to establish the identity of the person in a pre-contract meeting, and make sure this is recorded in writing.

Clerk of works

5.5 The employer is entitled to employ an independent clerk of works whose duty is 'to act solely as an inspector on behalf of the Employer under the direction of the Architect' (cl 3.10). The presence of a clerk of works does not lessen the architect's own duty in respect of site inspection *(Kensington Health Authority v Wettern)* and the architect should not treat the clerk of works as an agent carrying out work on behalf of the architect. Unlike JCT98, IFC98 does not give the clerk of works any authority to issue directions, therefore the role is confined to inspection on behalf of the employer. It would be implied that the contractor should give the clerk of works reasonable access and facilities, but it might be sensible to make this clear in the tender documents, particularly if the clerk of works is to maintain a permanent presence on site.

Kensington and Chelsea and Westminster Area Health Authority v Wettern Composites (1984) 31 BLR 57

Wettern Composites were sub-contractors for the supply and erection of pre-cast concrete mullions for an extension to the Westminster Hospital, on which the health authority had also engaged architects, engineers and a clerk of works. Tersons Ltd were the main contractors. The hospital was completed in 1965 and in 1976 it was discovered that there were considerable defects in the mullions. The health authority brought an action against the architects, engineers and sub-contractors, though the latter subsequently went into liquidation. Judgment was given for the health authority. The architects had failed to exercise reasonable skill and care in ensuring conformity of the works to the design. Although a clerk of works was employed this did not lessen the architect's responsibility. However, the health authority was vicariously liable for the contributory negligence of their clerk of works, and the damages recoverable from the architects were reduced by 20 per cent accordingly.

Planning supervisor

5.6   The planning supervisor's duties derive from the CDM Regulations, and it is the employer's obligation under the contract to ensure that the planning supervisor complies with these (cl 5.7.1). It is the contractor's responsibility to develop the Health and Safety Plan so that it complies with the CDM Regulations, and to ensure that the Works are carried out in accordance with the Plan. The planning supervisor will monitor the development of the Plan, but has no duty to inspect the Works and would be very unlikely to visit the site unless there is some very unusual circumstance, such as the discovery of an unanticipated hazard. The main responsibility for ensuring correct health and safety measures are employed on site rests with the contractor, both under statute and under the express terms of the contract (cl 5.7.2).

Information to be provided by the architect

5.7   One of the key duties of the architect under a construction contract is to supply the contractor with sufficient information to construct the Works in accordance with its terms. In an ideal world the contractor should be supplied with every piece of information required at the very start of the project but in practice this ideal is rarely achieved. Even if the Works have been fully specified it is likely, eg, that information regarding assembly, location, detail dimensions, colours, etc will be needed by the contractor throughout the project. In many cases it will not have been possible to prepare this in advance as detailed information regarding the site, or perhaps in relation to named sub-contractor design items, will not have been available. In addition, circumstances may result in the need for a variation and the provision of revised or entirely new information to the contractor. Supply of information will usually form part of the architect's express or implied duties to the employer under the terms of appointment.

5.8   IFC98 refers in three places to the architect's obligation to provide information. These refer to setting out information (cl 3.9): 'information referred to in the information release schedule' (cl 1.7.1); and 'such further drawings or details which are reasonably necessary to explain and amplify the Contract Drawings' (cl 1.7.2). Although none of the information is required to be released under an 'architect's instruction', this is frequent practice and wise as it would enable the clause provisions to be brought into operation if necessary *(see Architect's instructions, para 5.19 below)*. If any of the information supplied introduces changes or additions to the Works it must be covered by an architect's instruction requiring a variation.

5.9   Under clause 3.9 the architect is responsible for giving sufficient 'accurately dimensioned drawings' and for determining levels to enable the contractor to set out the Works. The contractor must at no cost to the employer 'amend any errors' that result from its own inaccurate setting out. Alternatively, the architect, with the employer's consent, may instruct that the error remains, in which case 'an appropriate deduction ... shall be made from the contract sum' (cl 3.9). There is no suggestion in the Conditions as to how this might be assessed. In practice it will be a matter for negotiation, based on the anticipated losses to the employer through, eg, the resulting reduction in value of the property plus any costs in professional fees for redesign. The error and the deduction should first be discussed with the employer, and the agreed deduction should ensure adequate compensation.

5.10   Information shown on the information release schedule must be supplied at the stipulated date. Failure to provide the information covered by clause 1.7.1 causes delay, an event for which an extension of time may be granted (cl 2.4.7.1) and may give rise to a direct loss and/or expense claim where the failure causes such loss or expense (cl 4.11). It could also be grounds for determination, but only where such failure has led to the suspension of the carrying out of the whole of the Works for a continuous period of one month (cl 7.9.2(a)). The obligation is qualified by the proviso 'except to the extent that the Architect is prevented by the act or default of the Contractor or of any person for whom the Contractor is responsible'. This might apply where, eg, completion of services drawings was dependent on the contractor having dug trial holes to establish existing drainage routes, and the contractor failed to carry this out.

5.11   The contract allows the employer and contractor to agree changes to the information release schedule (cl 1.7.1). There is no mechanism whereby the architect may unilaterally adjust the schedule if the contractor falls behind programme. It should be noted that a extension of time would not be automatic simply because a drawing is provided late, the contractor would still have to show that the late information had caused a delay, and if the contractor is behind programme this may be difficult. It would therefore be in both parties' interests to negotiate adjustments to the schedule, otherwise in time it will bear little relation to the real programme of works. If variations have been issued that involve additional work and have resulted in an extension of time, then again it would be reasonable for the schedule to be adjusted to reflect this. Care should be taken by the architect to point out if any agreed acceleration is likely to present difficulties for his or her own programming – the architect is unlikely to be found liable for failing to meet this accelerated programme if the client was warned at an early stage that it may be unachievable.

5.12   With respect to information not shown on the schedule, or where a schedule is not used, the architect is under an obligation to provide 'such further drawings and details as are reasonably necessary either to explain and amplify the Contract Drawings ... and shall issue such instructions to enable the contractor to carry out and complete the Works' (cl 1.7.2). The information and instruction should be provided in sufficient time to allow the contractor to complete by the Date for Completion, or, if the contractor appears unlikely to complete by this date, at a date when 'having regard to the progress of the Works' it is reasonably necessary for the contractor to receive the information.

5.13   If the contractor is aware and has 'reasonable grounds for believing the architect is not so aware' of when information may be needed the contractor should advise the architect (cl 1.7.2). It should be noted that there is no requirement for this to be done in writing, therefore the architect should make careful notes of any conversations, eg on-site visits, where information is requested. Failure to provide the information and instructions under clause 1.7.2 is a clause 2.4 event, a 'matter' which may give rise to a direct loss and/or expense claim (cl 4.12.1.2), and possible grounds for determination as discussed above. Consultants are not referred to in IFC98, but delay in supplying drawings by, say, an engineer would have the same effect under the contract as a delay on the part of the architect, ie a delay for which the employer is responsible.

Information provided by the contractor or named sub-contractor

5.14   The contractor as 'Principal Contractor' may be required by the planning supervisor to provide information in relation to the Health and Safety File (cl 5.7.4). It should be noted, however, that IFC98 does not contain express provisions for 'as built' drawings. If these are needed the specific requirement must be set out in the Bills and specification.

5.15   If ESA/1 is executed, the named sub-contractor will have an obligation to provide information to the architect in accordance with any time requirements set out in Part 3 of the Appendix to the form (ESA/1 cl 3). This information is stated under clause 3 to be for the purposes of either obtaining the main contract tenders or instructing the expenditure of a provisional sum relating to the named sub-contractor, or to enable the architect to issue information to either the main contractor or the specialist.

5.16   In addition, clause 2 states that 'the information required to be provided ... shall also be provided so as to enable the Architect to coordinate and integrate the design of the Sub-Contract Works into the design for the Main Contract Works as a whole'. It is suggested that the combined effect of these two clauses is not entirely clear. However, it appears that as clause 3 refers only to information provided before the sub-contract is entered into, the requirement in clause 2 is similarly confined to this information. In other words, none of these obligations relate to additional information which might be needed after the sub-contractor has been named. The best route in practice would be to ensure that all information had been received before the sub-contractor is named, otherwise the terms of ESA/1 may need to be amended. It would also appear that the clause 2 requirement could not be relied on to request information any earlier than the dates given in the Appendix to ESA/1.

5.17   The sub-contractor would be directly liable to the employer for any breach of the obligation to provide information, and the employer would therefore be able to bring a claim against the named sub-contractor for any losses suffered. It is notable that there are no terms in the main contract which expressly relieve the main contractor from liability, but as there are no terms requiring the contractor to provide information, is is very unlikely that the employer could claim losses from the main contractor. Any risk of late named sub-contractor information holding up progress would be borne by the employer through the operation of the extension of time and disruption provisions.

5.18   This named sub-contractor's obligation to provide information is subject to the employer providing information according to any dates agreed under Appendix 2 to ESA/1. This means that if the information is not provided according to these dates, the employer will not be able to bring any claim against the named sub-contractor for losses suffered. If calendar dates are agreed, there are no means of adjusting them should the main contractor's (and consequently the sub-contractor's) programme be delayed. In practice, therefore, it may be better to relate those dates to work stages rather than calendar dates.

Architect's instructions

5.19   Only the architect has the power to issue instructions. Sometimes the architect may issue instructions (eg instructions requiring a variation under clause 3.6), but at other times the architect shall issue instructions (eg instructions regarding discrepancies between contract documents under clause 1.4). The latter is an obligation. If the employer gives an instruction other than through the architect this would not be effective under the contract. The contractor would be under no obligation to comply with any such instruction. If the contractor, however, does carry out the instruction a court might decide either that there had been an agreed amendment to the contract, or that the instructed work is not part of the contract but a separate agreement between the contractor and the employer. The consequences of such an agreement would be difficult to sort out in practice and the employer would be very unwise to make such agreements or issue any instructions other than through the architect.

5.20   Where the architect acts outside his or her authority, the contractor would be under no obligation to comply with the instruction given. On the other hand, provided the architect is acting within the terms of the contract, the contractor must comply, even if the architect's action is contrary to the express requirements or instructions of the employer.

5.21   Clause 3.5.1 states that all instructions must be in writing. The provisions make no reference to oral instructions, which would therefore be of no effect. The contractor would not be obliged to comply with any oral instruction and would be wise to request all instructions to be confirmed immediately in writing before taking action. If the contractor carries out a variation on the basis of an oral instruction only, then the architect could later sanction the instruction in writing (cl 3.6), but the contractor would be taking a risk. The contract contains no provisions to cover the situation where the contractor confirms an oral instruction in writing; however, it might be wise to respond promptly reminding the contractor of the correct procedures under the contract.

5.22   There is no special format required for instructions, but it is often convenient to use the forms published by RIBA Publications. An instruction in a letter would be effective, even if the normal practice on that project was to use the printed forms, as long as the letter is quite clear. A drawing sent with a letter requiring it to be executed would constitute an instruction, but a drawing with no covering instruction may be ineffective. All instructions must be given to the contractor, or the contractor's agent on site, even if the instructions relate to named sub-contractor's work.

5.23   Instructions in site meeting minutes may constitute a written instruction if issued by the architect, but not if issued by the contractor, and only if the minutes are recorded as agreed at a subsequent meeting. It is possible that the instruction would not take effect until after the minutes were agreed, and it would depend on the circumstances whether the minutes were sufficiently clear to fall within the terms of the contract. It is therefore not good practice to rely on this method. Site instruction books should also be avoided. Signing an instruction in a book would constitute a written instruction under the terms of the contract, but there is no obligation to sign such books, and it may be prudent not to make quick decisions on site but to wait until all implications can be checked. With the possibility of faxed instructions the delay should be very short.

5.24   The contractor must comply with every instruction **(see figure 8)** provided that it is valid, ie provided that it is in respect of a matter regarding which the architect is empowered to issue instructions **(cl 3.5.1)**. The contractor must 'forthwith' comply, which for practical purposes means as soon as is reasonably possible except that the contractor need not comply with a clause 3.6.2 instruction (access and use of the site, etc) to the extent that it makes a reasonable objection **(cl 3.5.1)**.

5.25   If the contractor feels that an architect's instruction might not be empowered by the contract, or requires clarification, then the contractor may ask the architect to specify in writing the provisions of the contract under which the instruction is given, and the architect must do this 'forthwith' **(cl 3.5.2)**. (To avoid this happening, it is good practice to always name the clause under which the instruction is empowered.) The contractor must then either comply or issue a notice referring the disputed instruction to the decision of an adjudicator. If, however, the contractor chooses to accept the architect's reply and complies with the instruction, then the employer is bound by the instruction. This would appear to be the case even if at a later stage it is established that the architect had no authority under the contract.

Figure 8 Key matters about which the architect is empowered to issue instructions

```
 – divergence (cl 1.4);
 – making good defects (cl 2.10);
 – name person as sub-contractors (cl 3.3.2);
 – variations (cl 3.6);
 – expenditure of provisional sums (cl 3.8);
 – setting out (cl 3.9);
 – opening up and tests (cl 3.12);
 – removal of defective work (cl 3.14.1);
 – work  not carried out in a proper and workmanlike manner (cl 3.14.2);
 – postponement (cl 3.15);
 – statutory requirements (cl 5.2);
 – insurance (cl 6.2.4, 6.3D.1).
```

5.26   Even if the contractor decides to query the instruction under clause 3.5.2, this does not relieve the contractor from the obligation to comply. Should the instruction be found to be valid the contractor would be liable for any delay caused by failing to comply as required by the contract. If the contractor does comply, but the instruction turns out to have been invalid, the contractor may be entitled to any losses incurred through compliance. The contractor would have to make a commercial decision regarding whether to comply or await the outcome, but the architect would be wise to deal promptly with any such query.

5.27   If the contractor does not comply with a written instruction the employer may employ and pay others to carry out the work to the extent necessary to give effect to the instruction (cl 3.5.1). The architect must have given written notice to the contractor requiring compliance with the instruction, and seven days must have elapsed after the contractor's receipt of the notice before the employer may bring in others. This suggests that some recorded form of delivery is desirable. Although there is no obligation to issue such notices, it would prudent to take swift action in order to protect the employer's interests. The employer is entitled to recover any additional costs from the contractor, ie the difference between what would have been paid to the contractor for the instructed work, and the costs actually incurred by the employer. These costs could include not only the carrying out of the instructed work, but any special site provisions that would need to be made, including health and safety provisions, and any additional professional fees charged. Although it would be wise to obtain alternative estimates for all these costs wherever possible, if the work is needed urgently there would be no need to do so.

Variations

5.28 Architect's instructions often require some variation to the Works. Under common law neither party to a contract has the power unilaterally to alter any of its terms. Therefore, in a construction contract neither the employer nor architect would have the power to require any variations unless the contract contains such a power. As some aspects of construction may be difficult to define exactly in advance, most construction contracts contain provisions allowing the employer to vary the Works to some degree. Changes can arise because of unexpected site problems, or because of design changes wanted by the employer, or because the architect has to change information issued to the contractor.

5.29 Under IFC98 the architect is empowered to order specific variations (cl 3.6). The power is broadly defined and includes alterations to the design, quality and quantity of the Works, and to operational restrictions such as access to the site. The contract expressly states that no variation will vitiate the contract (cl 3.6) but the power does not extend to altering the nature of the contract, nor can the architect issue variations after practical completion. All variations under clause 3.6 may result in an adjustment of the contract sum (cl 3.7.1) and give rise to a claim for an extension of time (cl 2.4.5), or for direct loss and/or expense (cl 4.12.7). If the works are suspended for a period of one month or more as a consequence of the variation, this would be grounds for the contractor to determine its employment under the contract, unless the variation was necessitated by some negligence or default of the contractor (cl 7.9.2).

5.30 The architect may vary the Works, eg by changing the standard of a material specified. The architect may add to or omit work, or substitute one type of work for another or remove work already carried out (cl 3.6.1). The architect may vary the access to or use of the site, limitations on working space or working hours, the order in which the work is to be carried out, or any restrictions already imposed (cl 3.6.2) (SMM7 requires these to be set out separately). However, the contractor need not comply to the extent that it makes reasonable objection (cl 3.5.1). Given that the contractor will be paid for such variations it is difficult to see what might constitute a 'reasonable' objection but, eg, a variation might have a detrimental 'knock-on' effect on some other project causing the contractor to suffer losses for which it would not otherwise be compensated or an instruction that would make site operations almost impossible to manage. This contract provision is necessary not only to allow the employer some flexibility, but also to accommodate difficulties that may arise, eg, through local authority restrictions on working hours.

5.31   The architect may issue an instruction to postpone work (cl 3.15). Any such instruction may give rise to a claim for an extension of time (cl 2.4.4) or direct loss and/or expense (cl 4.12.5), and if the postponement results in a suspension of work for a period greater than that stated in the Appendix, then the contractor would have grounds for determination. The consequences of such an instruction are therefore serious, and the architect would advise any client who is suggesting such a measure accordingly. It is difficult to envisage situations where it would be necessary to postpone work, but it could arise where there have been problems with statutory approval, or reaching agreement over a boundary matter, and the only option might be to postpone the relevant part of the Works.

5.32   Finally, the architect may sanction any variation made by the contractor other than under an instruction of the architect (cl 3.6). If such a variation were likely to affect the employer, the architect would be wise to discuss it with the employer before taking action.

Goods, materials and workmanship

5.33   Clause 1.1 makes it clear that all work must be carried out in accordance with the standard specified in the contract documents. The architect will normally inspect at regular intervals to monitor the standard that is being achieved. If any changes were made in order to raise or lower the standard then this would constitute variation. When the standard achieved appears to be unsatisfactory it can be tempting to become involved in directing the day-to-day activities of the contractor on site. Apart from being an enormous burden on the architect, this could confuse the issue of who is ultimately responsible for quality and is to be avoided. The architect would normally, of course, draw the contractor's attention to areas of defective or poor quality work. There are also some measures set out in the contract.

Defective work

5.34   The architect may instruct the contractor to open up completed work for inspection, or arrange for testing of any of the work or materials, fixed or unfixed (cl 3.12). Obviously the architect would only do this if there are reasonable grounds for suspecting defective work or materials. No time limit is specified, but obviously the architect should instruct as soon as the need for such action becomes apparent (delay could result in escalating or unnecessary costs), although failure to ask for tests in no way relieves the contractor from the obligation to provide work according to the contract. The architect should explain to the employer the need for the tests and their contractual implications. The cost of carrying out the tests is added to the contract sum, unless it was already provided for in the Bills of Quantities under a provisional sum, or unless the work proves to be defective. Unless the work is defective the contractor may also be entitled to an extension of time under clause 2.4.6 and loss and/or expense under clause 4.12.2.

5.35   If work is found to be defective, the contractor must write to the architect proposing action to be taken immediately to establish whether there are any similar problems in work already carried out. If the architect has not received proposals within seven days, or is not satisfied with the contractor's proposals, or cannot wait seven days to receive the proposals, the architect can instruct that further tests or opening up be carried out (cl 3.13.1). In this case the cost of further tests would be borne by the contractor whether or not the additional tests demonstrated work to be defective. The contractor has a right of objection, which must be made within ten days of receipt of the instruction (cl 3.13.2), but whether or not he exercises this right it must comply with the instruction immediately. Following the objection the architect may either withdraw his or her instruction or modify it – if neither is done within seven days of the objection, the contractor would have to take the dispute to adjudication.

5.36   The architect has the power to issue an instruction requiring the removal of work, materials or goods from the site (cl 3.14.1). Even though this might appear rather excessive considering that the contractor is already under an obligation to build the work correctly, it can be important to issue such instructions as they enable the clause 3.5.1 sanctions to be brought into operation. To fall under clause 3.14.1 the instruction must specifically require removal of the work from site, however impractical. Simply drawing attention to the defective work would not be sufficient *(Holland Hannen v Welsh Health Technical Services)*.

Holland Hannen & Cubitts (Northern) Ltd v Welsh Health Technical Services Organisation (1985) 35 BLR 1 (CA)
Cubitts were employed by the Welsh Health Technical Services Organisation (WHTSO) to construct two hospitals at Rhyl and Gurnos. Percy Thomas (PTP) were the architects. Redpath Dorman Long Ltd (RDP) were nominated sub-contractors for the design and supply of pre-cast concrete floor slabs. RDP assured WHTSO that the floors would be designed to CP 116 (concerning deflection), but the design team later required RDP to work to CP 204. Following installation, the contractor complained about extra work and costs due to adjustments to the partitions necessitated by excessive deflection of the

floors, and it was established that they had been designed to CP 116 not CP 204. PTP sent three letters 'condemning' the floors, but the first did not mention clause 6(4), and none of them required removal of the work. Cubitts stopped work for 20 weeks until PTP issued instructions specifying how the defect should be resolved. Cubitts commenced proceedings claiming compensation for delay. The claim was settled, but the relevant parties maintained their proceedings against each other for contribution. The Official Referee decided that RDL was liable for two-thirds of the amount paid to Cubitts and the design team for one-third. The Court of Appeal decided that this was incorrect and the correct apportionment should have been for RDL to be liable for one-third and the design team for two-thirds. In reaching this conclusion it stated: 'PTP contributed very substantially to the delay which occurred, in failing to recognise the defect in the design at an earlier stage; by issuing an invalid notice in 1976, and by moving very slowly thereafter to take the necessary steps to have the defects in the flooring put right' (Robert Goff LJ).

5.37   There are no express provisions in IFC98 whereby the defective work, materials or goods can be allowed to remain – if this is required it would need to be covered by an instruction authorising a variation. It is essential to secure the employer's consent and agree a reasonable reduction in the contract sum with both parties. The architect must specify in writing exactly which work may remain and record the agreed deduction. If variations to other work become necessary as a consequence it should be agreed (again following consultation with the employer and contractor) that no addition is made to the contract sum and no extension of time or direct loss and/or expense is given in respect of this. The architect should strongly advise the employer against accepting any defective work that could later cause technical problems or be a source of irritation. The difficult case of *Ruxley Electronics v Forsyth* illustrates that it may not be possible to claim the cost of having the work rebuilt at a later date.

Ruxley Electronics and Construction Ltd v Forsyth (1995) 73 BLR 1
Mr Forsyth employed Ruxley Electronics to build a swimming pool. The drawings and specification required the pool to be seven foot six inches deep at its deepest point, but the completed pool was only six foot nine inches deep. The contractors brought a claim for their unpaid account, and Mr Forsyth counterclaimed the cost of rebuilding the pool, which would be £21,560. The trial judge found that the shortfall in depth did not decrease the value of the pool and that Mr Forsyth had no intention of building a new pool. He rejected the counterclaim but awarded £2,500 as general damages for loss of pleasure and amenity. Mr Forsyth appealed and the Court of Appeal allowed the appeal and awarded him £21,500. The contractor appealed and the House of Lords restored the original ruling, confirming that the cost of reinstatement is not the only possible measure of damages for

defective performance of a building contract and is not the appropriate measure where the expenditure would be out of all proportion to the benefit to be obtained.

5.38   Clause 1.1 requires the contractor to carry out the work 'in a proper and workmanlike manner' and in accordance with the Health and Safety Plan. Clause 3.14.2 states that in the event of any failure, the architect may issue instructions requiring compliance, and these will not result in any addition to the contract sum, nor will they entitle the contractor to any extension of time or direct loss and/or expense. The clause empowers the architect to intervene in the contractor's working methods if necessary.

Sub-contracted work
5.39   IFC98 provides for two methods of sub-contracting work, both allowing for some control over which firms the contractor uses.

Domestic sub-contractors
5.40   Under clause 3.2 the contractor may only sub-contract work with the written consent of the architect. Failure to obtain this would be a default, providing grounds for determination under clause 7.2.1(d). Clause 3.2 states, however, that the architect's permission cannot be unreasonably withheld. It is suggested that permission is required for each instance of sub-letting, rather than agreeing to sub-letting in principle. There is no requirement to use a particular form of sub-contract though IN/SC has been developed for use with IFC98. Whatever form of domestic sub-contract is used, however, it must include certain conditions, and clause 7.2 states the sub-contract must provide that:
- the sub-contract is determined immediately upon determination of the main contract (cl 3.2.1);
- unfixed materials and goods placed on the site by the sub-contractor shall not be removed without written consent by the contractor (cl 3.2.2(a));
- it shall be accepted that materials or goods included in an interim certificate that have been paid by the employer become the property of the employer (cl 3.2.2(b));
- it shall be accepted that any materials or goods paid for by the main contractor prior to being included in a certificate become the property of the main contractor (cl 3.2.2(a));
- the sub-contractor has a right to interest on late payments by the contractor at the same rate as that due on main contract payments (cl 3.2.3).

5.41   These clauses protect the position of the employer and the provisions regarding unfixed goods and materials are of particular importance in this respect. If a main contractor should sub-contract on other terms, and this results in losses to the employer, then the contractor may be liable as this would be a breach of contract.

Named persons as domestic sub-contractors

5.42   The provisions for naming of sub-contractors are set out in clause 3.3. The sub-contractor can either be named in the specification/schedules/Bills, in which case the firm must have been selected prior to tendering the main contract, or in an instruction regarding a provisional sum. The latter route gives more flexibility to the employer but requires the employer to accept more risk.

5.43   Where the sub-contractor is named in the tender documents, the main contractor must be provided at the time of tender with NAM/T with Sections I and II completed, documents referred to as 'the Numbered Documents' that describe the work, which might include, for example, drawings and Bills of Quantities or specifications, together with a copy of the Health and Safety Plan (if and to the extent available), and a copy of the executed ESA/1 (if used). (A deed might be preferable where design work is involved, but there should be consistency between all documents.) Where the sub-contractor is named in an instruction relating to a provisional sum the same information must accompany the instruction.

5.44   In order to collate this information the architect must first invite sub-contract tenders using NAM/T. Section I is the invitation to tender, to be completed and signed by the architect, before it is sent to the invited sub-contractor. NAM/T Section I contains such important items as the 'dates between which it is expected that the sub-contract work will be carried out', insurance and VAT arrangements and dispute resolution matters. If ESA/1 is being used this should be completed and sent at the same time. The sub-contractor then completes NAM/T Section II and ESA/1 (in part – detailed guidance notes are given in the form), and returns both to the architect. If the offer is acceptable, the employer executes ESA/1, which then forms a binding agreement.

5.45   There is an alternative procedure for use in situations where there is insufficient information available to request a firm tender for the Works, but the employer wishes to enter into a warranty. ESA/1 can be sent on its own (in this case with paragraphs A and AA deleted) and the sub-contractor submits an approximate estimate for the Works. The warranty is executed on this basis, and a firm tender sought on NAM/T when more information is available.

5.46   Where the sub-contractor is named in the main contract documents, the contractor will price this work when submitting its tender. The contractor must then enter into an agreement with the named person within 21 days of entering into the main contract with the employer. The agreement is made using Section III of NAM/T, which refers to the sub-contract conditions NAM/SC (cl 3.3.1) (NAM/SC itself does not have to be executed). If the contractor is unable to enter into a sub-contract in accordance with the particulars in the main contract documents, the contractor must immediately inform the architect of the particulars that have prevented this from happening (cl 3.3.1) – note it does not state 'in writing'. The architect must then issue an instruction which could either 'change the particulars so as to remove the impediment' (cl 3.3.1(a)), or omit the work, substituting a provisional sum if wished (cl 3.3.1(b),(c)). It is suggested that, with the exception of cases where a named sub-contractor's employment has been determined *(see para 9.31 below)*, the main contractor could not be required to carry out work which the tender documents stated were to be undertaken by a named (or to be named) sub-contractor.

5.47   An instruction under clause 3.3.1(a) or (b) is to be treated as a variation under clause 3.6, which may give rise to a claim for an extension of time, and also be a matter with respect to a claim for direct loss and/or expense (although this would be unlikely where the work is omitted). An instruction under clause 3.3.1(c) is dealt with in accordance with clause 3.3.2, ie as an instruction relating to a provisional sum, and therefore could also give rise to an adjustment of the contract sum, and an award of an extension of time or direct loss and/or expense. In addition, as with any instructions, if the architect does not issue the necessary instructions promptly, this could give rise to claims under clauses 2.4.7 and 4.12.1 (delay or disruption due to failure to issue necessary instructions) and, where the Works are suspended by one month as a consequence, to determination by the contractor. It is suggested that this would only apply where the architect has failed to act within a reasonable time *(Percy Bilton Ltd v Greater London Council)*. It is clear, however, that the risk of a problem developing in finalising the sub-contract details is borne in part by the employer, and it may be prudent for the architect to check before the main contract tender is accepted that there appear to be no outstanding matters to be resolved before the sub-contract can be formed, particularly in cases where a considerable period has elapsed since the named person's tender was sought.

Percy Bilton Ltd v Greater London Council (1981) 20 BLR 1 (HL)
Percy Bilton contracted with the GLC for the provision of 182 dwellings. WJ Lowdell Ltd were nominated for mechanical services, but after commencing the work eventually withdrew its labour and went into liquidation. A second firm was then nominated, but this company withdrew its tender. Finally, a third firm was nominated and a sub-contract was entered into. The contractor claimed an extension of time which was only granted in part, and the contractor then brought proceedings for the return of liquidated damages that had been deducted. The GLC

claimed that the problems of re-nomination were covered in clause 23(g) (delay on the part of a nominated sub-contractor) and therefore the extension of time and the deduction of liquidated damages was valid. Percy Bilton claimed that clause 23(g) did not cover that situation which amounted to a breach on the part of the employer. The House of Lords found that clause 23(g) did not cover the situation, but that withdrawal alone did not constitute a breach by the employer. The risk of delays caused by withdrawal lay with the contractor. However, the employer had a duty to re-nominate within a reasonable time, and failure to do so could constitute a relevant event under clause 23(f) (failure to issue necessary instructions).

5.48   Where the sub-contractor is named in an instruction relating to a provisional sum, the contractor may make reasonable objection to the named sub-contractor, but must do so within 14 days of the date of issue of the instruction (cl 3.3.2). Otherwise, the contractor must enter into the sub-contract 'forthwith', using documents as described above. The contract does not say what will happen if an objection is made, if the objection is reasonable the architect would have to name another person in a further instruction, or issue instructions to remove the impediment as described above. If a dispute arises this could be referred to adjudication. The instruction naming the sub-contractor is to be valued under the rules in clause 3.7. Delays arising from compliance with the instruction would be grounds for an extension of time under clause 2.4.5 and loss/expense under clause 4.12.7, and as above, delays in issuing the instructions could give rise to claims under clauses 2.4.7 and 4.12.1, or even to determination.

5.49   Once the sub-contract is entered into, the contractor is entirely responsible for the work carried out by the sub-contractor, and delays on the part of the sub-contractor are not grounds for an extension of time or direct loss and/or expense. Responsibility for payment also rests entirely with the contractor, The architect is therefore not concerned with the details of the sub-contract terms, and the contractor is not required to send a copy of the executed NAM/T Section III to either the architect or the employer. If the architect would prefer to have a copy then the contract should be amended accordingly.

5.50   Under certain circumstances the architect may be obliged to name a replacement sub-contractor. This could occur if the named sub-contractor's employment is determined by the contractor or the named sub-contractor determines its own employment (cl 3.3 (a)). The determination procedures are discussed at paras 9.30–9.34 below.

**Work not forming part of the contract/
persons engaged by the employer**

5.51   Under clause 3.11 the employer may engage persons direct to carry out work
        that does not form part of the contract, while the main contractor is still in
        possession. This may include statutory undertakers when employed by the
        employer, but not where they are carrying out the work 'in pursuance of their
        statutory duties'. If the contract documents have included this requirement,
        then the contractor must permit the employer to execute such work. Otherwise
        the employer can only do this with the contractor's consent. The consent may
        not be unreasonably delayed or withheld.

5.52   Clause 3.11 makes it clear that for the purposes of insurance the contractor is
        not responsible for the directly engaged person. The employer should therefore
        ensure that insurance cover is arranged in respect of any act or neglect of the
        persons to be employed. The employer should also be made fully aware that
        any disruption caused to the contractor's working could lead to a claim for an
        extension of time (cl 2.4.8), to loss and expense (cl 4.12.3), or even to the
        contractor determining the contract (cl 7.9.2(c)). The employer is therefore at
        considerable risk, and should be advised to avoid this route if at all possible.

Making good defects

5.53   The contractor is required to make good any 'defects, shrinkages or other faults
        which appear and are notified by the Architect/the Contract Administrator to
        the Contractor' (cl 2.10). The defects are limited to those that result from the
        Works not having been carried out in accordance with the contract. This does
        not include other defects that may be due, eg, to errors in the design
        information supplied to the contractor, or to general wear and tear resulting
        from occupation by the employer, or shrinkages which would be expected even
        if the Works had been carried out as specified. The contractor is similarly not
        liable for frost damage that occurs after Practical Completion.

5.54   The obligation does not appear to be limited to those defects that appear after
        Practical Completion, and therefore could extend to defects that were patent
        at that time (note that the wording is different from that in JCT98). It is
        suggested that the obligation would include defects caused by frost occurring
        before Practical Completion.

5.55   However, it is important to note that the obligation to make good appears to
        be limited to those defects notified by the architect. It is therefore important
        that the architect prepares a comprehensive schedule. The architect should
        issue the notification not later than 14 days after the end of the defects liability

period, the only point where the contract requires the architect to issue such a schedule. The architect may require the contractor to make good a defect at an earlier stage, but this should only be used for serious and urgent problems. No special format is required for the notification, but it is common practice to issue it in the form of an instruction.

5.56   If the architect, with the agreement of the employer, decides to accept any defective work then this should be confirmed clearly in writing. Care should be taken to establish the full extent of the problem before such a course of action is taken, and an appropriate deduction from the contract sum agreed, as it is unlikely that the employer would thereafter be able to claim for consequential problems or further remedial work.

5.57   Once satisfied that all the notified defects have been made good, the architect must issue a certificate to that effect (cl 2.10). The certificate is one of the pre-conditions to the issue of the Final Certificate. Note that a separate certificate will be needed for the 'relevant part' if the partial possession provisions have been operated (cl 2.11), and that separate certificates will be needed for each section if the Sectional Completion Supplement has been used.

5.58   The contract does not state what should happen in respect of defects which appear after the issue of the certificate but before the issue of the Final Certificate. It is, however, clear from clause 2.10 that the architect no longer has the power to instruct that these are made good. It is suggested that in such circumstances there would be two possible courses of action. The first would be to make an agreement with the contractor to rectify the defects before the Final Certificate is issued. If the contractor refused to do this an amount could be deducted from the contract sum to cover the cost of making good the work, but this would involve some risk to the employer. The second and less risky course would be to have the defective work rectified by another contractor, and deduct the amount paid from the contract sum. This would involve a delay in the issue of the Final Certificate and would probably be disputed by the contractor.

5.59   The contractor's liability for defective work does not end with the Final Certificate, except to the limited extent in which the Final Certificate is conclusive. The contractor is still liable for losses suffered, but no longer has the right to return to site to correct defective work. The employer's remedy is to bring an action at common law. The defects liability period therefore is a sensible procedure which benefits the parties in affording an opportunity to remedy problems at a reasonable cost, without the problems associated with bringing a legal action. Nevertheless if the architect fails to notify the contractor of patent defects, although the contractor would still be liable for any defects, if particular difficulty is experienced in getting these remedied at a later date, then the employer may look to the architect for compensation for any losses.

6.1   The contract sum, which will be the tender figure accepted or agreed following negotiation, is entered in Article 2. However, the wording of the contract recognises that in practice this is rarely the amount actually paid, and refers in Article 2 to the contract sum 'or such other sum as shall become payable'. It should be noted that the contract sum itself does not change, the provisions refer to adding or subtracting amounts from the contract sum to reach an adjusted figure.

6.2   There are many reasons why the amount finally payable will differ from the contract sum. Under IFC98 the sum may contain provisional sums or approximate quantities to cover the cost of work that cannot be accurately described or measured until work is under way. If variations to the Works are instructed, then the amount payable must be adjusted accordingly. There is also the possibility of claims from the contractor for loss and expense arising from intervening events which could not be foreseen at the time of tendering. Fees or charges in respect of statutory matters which are not allowed for in the contract documents will require an adjustment to the sum. IFC98, like most contracts, contains 'fluctuations' provisions allowing for adjustments in the event of changes in statutory charges, or the market price of materials and labour. VAT is, of course, not included in the contract sum.

6.3   There will therefore almost inevitably be adjustments during the course of the contract; IFC98 clause 4.1, however, makes it clear that the only alterations that may be made are those provided for in the terms of the contract. Usually the ascertained amounts will be added or deducted as appropriate at the periods for certification. Arithmetical errors by the contractor in pricing are not allowed as a cause for adjustment. Errors in the preparation of the contract Bills, if used, must be corrected and will then be treated as if they are a variation (cl 1.4). Any divergence between the contract drawings and other documents which necessitates an instruction by the architect may also result in a variation (cl 1.4).

An approximate quantity

6.4   Where Bills of Quantities are used and work can be described in accordance with the Standard Method of Measurement 7th Edition (SMM7), but the quantity involved is uncertain, an 'Approximate Quantity' can be included. The architect is not required to issue any further instruction for the contractor to carry out this work. After it has been carried out, the work is valued under option A or option B as described in clause 3.7.1 *(discussed at para 6.12 below)*. If the contractor wishes to submit a 'Price Statement' under option A this must be done within 21 days of the commencement of the work to which the approximate quantity relates (cl 3.7.1 A1).

6.5   Difficulties can arise if the approximate quantity is not a reasonably accurate forecast of the quantity of work required. The valuation must then include a fair allowance for the difference in quantity over and above the rates or prices tendered by the contractor (cl 3.7.4 (b)). The inaccuracy is also an event which may give rise to an extension of time under clause 2.4.15 and a 'matter' for direct loss or expense under clause 4.12.8.

Provisional sums

6.6   Where Bills of Quantities are used but it is not possible to provide sufficient information at the time of tender to allow an item to be described and measured according to the SMM7 rules, then a provisional sum may be inserted in the Bills to cover the item. The architect must issue instructions with regard to the expenditure of any provisional sums (cl 3.8) and the contractor can take no action until receiving an instruction. Under SMM7, provisional sums are either for defined work or for undefined work. The SMM7 definitions were helpfully set out in the contract in footnote [bb] to clause 8.3, although this footnote has now been removed by Amendment 3 to the form.

Defined work

6.7   The information required to place provisional work in the defined category is listed in SMM7 General Rule 10.3. The tenderer must be aware of the nature and construction of the work, how and where the work fits into the building, the scope and extent of the work and any specific limitations on method or sequence or timing. In other words, the description must be sufficiently detailed for the contractor to make proper allowance for the effect of the work when pricing the relevant preliminaries, and to allow for the work in the programme.

6.8   If the information provided is not as detailed as the Rule requires, or if it is erroneous, then a corrective instruction is required from the architect. This will be deemed a variation (cl 2.2.2.2) and could give rise to a notice of delay and an application for reimbursement of direct loss and expense from the contractor. The corrective action cannot simply be to change from the defined category to the undefined category by substituting a new provisional sum.

Undefined work

6.9   A provisional sum for undefined work will be applicable where it is not possible to supply the amount of information needed to comply with Rule 10.3. The contractor will not have been able to make proper allowance for the work in programming, planning, or the pricing of preliminaries. A provisional sum in this case should be sufficient not only to cover the net cost, but also to take into account the fact that there might be additions to preliminaries, attendance, and perhaps loss and expense, etc. There is also a risk that the contractor might give notice of delay arising from the architect's instruction. Unlike an instruction for the expenditure of a provisional sums for defined work, that for a provisional sum for undefined work could be an 'event' (cl 2.4.5) and a 'matter' for which a loss and expense application can be made (cl 4.12.7).

6.10   Provisional sums may be included for items that are not specifically work, for example testing, site boards, site facilities, etc. The heading in the Bills of Quantities will simply be 'include the Provisional Sum of _____ for _____', or some other appropriate wording. For work to be carried out by statutory authorities it is suggested that the description of the work be followed by a similar heading.

Instructions on provisional sums

6.11   The architect must issue instructions regarding all work covered by provisional sums (cl 3.8). The work is then valued under option A or option B as described below for a variation.

Valuation of variations

6.12   There are two mechanisms by which a variation can be valued under the provisions of the contract the 'Contractor's Price Statement' ('Option A'); or valuation by the quantity surveyor ('Option B'). Of course it is always open for the contractor and employer simply to agree a price.

Option A: contractor's price statement

6.13   The text of 3.7 includes a provision (option A) whereby on receiving an instruction the contractor may on its own initiative submit a 'Price Statement' in respect of variations, provisional sum work and work covered by an approximate quantity. The contractor must submit the price statement to the quantity surveyor within 21 days of receiving the instruction or of receiving sufficient information to prepare the statement, if later. The statement should state the price, which must be based on the valuation rules set out in clause 3.7.2 to 3.7.10 *(see below)*, and may also attach the contractor's requirements for direct loss and/or expense and any extension of time. NAM/SC contains similar provisions whereby the sub-contractor may submit a statement to the contractor within 17 days of receiving an instruction requiring a variation (NAM/SC cl 16.2).

6.14   The quantity surveyor, after consultation with the architect, must notify the contractor within 21 days whether or not the price statement is accepted (cl 3.7.1.2A2). Where the statement is not accepted by the employer the quantity surveyor must notify the contractor of the reasons, set out in similar detail to the statement, and include an amended statement. The contractor must state within 14 days whether or not it accepts the amended price statement, if not either party may refer the matter to adjudication. If the matter is not referred to adjudication then the instructions are valued by the quantity surveyor under option B. A similar process is followed with respect to any claims for direct loss and/or expense and extensions of time attached to the statement, which can be accepted or rejected independently of the statement. Again, the quantity surveyor is required to consult with the architect (cl 3.7.1.2A7). Although the clauses do not require it, the contractor should prepare the claim on the basis of the terms of the contract, and the architect should assess the claim carefully before reaching a decision.

Option B: valuation by the quantity surveyor

6.15   If the contractor does not exercise the option to submit a price statement, or the price statement is unacceptable, then the variation must be valued by the quantity surveyor. The valuation rules to be used are set out in clause 3.7.2 to 3.7.10.

6.16   Omissions are valued in accordance with the values in the 'priced document' (the priced specification, schedules of work or contract Bills, or the contract sum analysis or the schedule of rates) (cl 3.7.3). Work of 'similar character' to that in the priced document is valued according to prices stated in that document, with a fair allowance being made if the conditions change or the quantity changes significantly (cl 3.7.4(a)). Dissimilar conditions might include, eg, that the instructed work is carried out in winter, whereas under the Bills it had been assumed it would be carried out in summer. Such an assumption, however, would have to be clear from an objective analysis of the contract documents *(Wates Construction v Bredero Fleet)*.

Wates Construction (South) Ltd v Bredero Fleet Ltd (1993) 63 BLR 128
Wates Construction entered into a contract on JCT80 to build a shopping centre for Bredero. Some sub-structural work differed from that shown on the drawings and disputes arose regarding the valuation of the works, which were taken to arbitration. In establishing the conditions under which, according to the contract, it had been assumed the work would be carried out, the arbitrator took into account pre-tender negotiations and the actual knowledge that Wates gained as a result of the negotiations, including proposals that had been put forward at that time. Wates appealed and the court found that the arbitrator had erred by taking this extrinsic information into consideration. The conditions under which the works had to be executed had to be derived from the

express provisions of the Bills, drawings and other contract documents.

6.17   Work for which an approximate quantity is included in the contract documents is valued at the stated rates and prices, except that where the approximate quantity was not a reasonably accurate forecast of the amount of work carried out, a fair allowance for the difference is to be included (cl 3.7.4(b)).

6.18   Work not of similar character should be valued at 'fair rates and prices' (cl 3.7.5). Where the appropriate basis of a fair valuation is Daywork, clause 3.7.6 sets out the rules to be followed in assessing the amount. Otherwise it would up to the quantity surveyor in the first instance to determine what a fair rate might be. If the contractor disagrees with this assessment, and the dispute cannot be resolved amicably, then the contractor could take the matter to arbitration.

Reimbursement of direct loss and/or expense

6.19   The objective of clause 4.11 is to enable the contractor to be reimbursed for direct loss and/or direct expense suffered as a result of deferment of possession or disruption to progress, and for which the contractor is not reimbursed under any other provision in the contract. As an alternative the contractor may be able to pursue this as a claim for general damages for breach of contract at common law, but this would need to be done through adjudication, arbitration or litigation.

6.20   The contractor must apply in writing, otherwise the architect has no power or obligation to deal with loss and expense. The application must be submitted promptly, in fact the use of the phrase 'is likely to incur' suggests an anticipatory quality. No particular format is specified but the application should be supported by such detail as is reasonably necessary, including identifying all 'matters' concerned *(London Borough of Merton v Leach)*. The architect or quantity surveyor can request additional information.

London Borough of Merton v Stanley Hugh Leach Ltd (1985) 32 BLR 51 (Ch D)
Stanley Hugh Leach entered into a contract with the council to construct 287 dwellings. The contract was JCT63. There was a considerable delay on the contract and a dispute arose regarding this delay and related claims for loss and expense. The dispute went to arbitration and the arbitrator made an interim award on a list of matters. Merton appealed and the court considered 15 questions framed as preliminary issues. Among other things the court stated that applications for direct loss and/or expense must be made in sufficient detail to enable the

architect to form an opinion as to whether there is any loss and/or expense to be ascertained. If there is, then it is the responsibility of the architect to obtain enough information to reach a decision. This could of course include requesting information from the contractor. The court also held that the application must be made within a reasonable time and not so late that the architect can no longer form an opinion on matters relevant to the application.

6.21   Claims can only be made for loss and expense suffered through deferment of possession, or the particular 'matters' listed in clause 4.12. Other losses are irrecoverable under the contract, although disputed claims may be referred to adjudication, arbitration or litigation. The matters listed in clause 4.12 are concerned with situations where the loss or expense is attributable to the employer, and excludes the 'neutral causes' which feature in the extension of time provisions of clause 2.4. Clause 4.11 refers to two types of claim, often difficult to separate because they are perhaps based on the same facts. The phrase 'regular progress of the Works being materially affected' allows for disruption not anticipated when tendering, and prolongation for which an extension of time may have been awarded. Any disruption should be related to the progress necessary to complete by the Date for Completion, not necessarily the actual sequences of events on site.

6.22   The sums that can be awarded can include any loss or expense that has arisen directly as the result of the 'matter'. The loss and expense award is in effect an award of damages and the architect should approach its assessment on the same principles as a court would when awarding damages for breach of contract. In broad terms, the object of the award is to put the contractor back into the position in which it would have been but for the disturbance. The contractor ought to be able to show that it has taken reasonable steps to mitigate its loss, and the losses must have been reasonably forseeable as likely to result from the 'matter' at the time the contract was entered into.

6.23   The following are items which could be included:
- increased preliminaries;
- overheads;
- loss of profit;
- uneconomic working;
- increases due to inflation; and
- interest or finance charges.

6.24   The items claimed must be things which the contractor could not recover under any other term of the contract (eg, it must not duplicate a claim under clause

3.7). Prolongation costs such as on-site overheads would normally only be claimable for periods following the Date for Completion. (For head office overheads, etc, see *McAlpine v Property and Land Contractors.*) Interest may also be recoverable, but only if it can be proved to have been a genuine loss *(FG Minter v WHTSO).* As clause 4.11 refers to losses which are 'likely to occur', the award need not be restricted to losses suffered prior to the time the contractor's application is made, but could include those suffered up until the date of the ascertainment (this would apply particularly to financing charges), and could arguably be extended to losses that could be predicted as likely to occur up until the date the reimbursement is made.

Alfred McAlpine Homes North Ltd v Property and Land Contractors Ltd (1995) 76 BLR 59
An appeal arose on a question of law arising out of an arbitrator's award regarding the basis for awarding direct loss and expense with respect to additional overheads and hire of small plant, following an instruction to postpone the works. The judgment contains useful guidance on the basis for awarding direct loss and expense. To 'ascertain' means to 'find out for certain'. It is not necessary to differentiate whether a head of claim is 'loss' or 'expense'. Regarding overheads, a contractor would normally be entitled to recover as a 'loss' the shortfall in the contribution that the volume of work had been expected to make to the fixed head office overheads, but which, because of a reduction in volume and revenue caused by the prolongation, was not in fact made. The fact that the 'Emden' or 'Hudson' formulae depend on certain assumptions mean that they are frequently inappropriate. The losses on the plant should be the true cost to the contractor, not based on notional or assumed hire charges.

FG Minter Ltd v Welsh Health Technical Services Organisation (1980) 13 BLR 1 (CA)
Minter was employed by Welsh Health Technical Services Organisation (WHTSO) under JCT63 to construct the University Hospital of Wales (second phase) Teaching Hospital. During the course of the contract several variations were made, and the progress of the works was affected due to the lack of necessary drawings and information. The contractor was paid amounts in respect of direct loss and/or expense, but the amounts paid were challenged as insufficient. The amounts had not been certified and paid until long after the losses had been incurred. It was argued that when there was a disruption to the works which resulted in the contractor having to lay out money which was not reimbursed through the normal interim valuations of the value of the works, then the contractor would suffer losses either through increased borrowing, or through using capital that would otherwise accrue interest. It was accepted by the court that, subject to the express terms of the contract, the contractor would normally be entitled to recover

such financing charges.

6.25   The contractor must provide full details and particulars of all items concerned with the alleged loss or expense. These should identify which of the losses claimed relate to each of the 'matters' that have occurred. This is sometimes compromised by the use of a 'rolled up' or composite claim approach where it is not really practicable to separate and itemise the effect of a number of causes. This has been accepted by the courts provided that as much detail as possible has been given, and provided that all disturbance was due to matters under clause 4.12, and not caused by the contractor.

6.26   Formulae such as the 'Hudson' or 'Emden' formulae are sometimes used for estimating head office overheads and profit, which may be difficult to substantiate. These can only be used where it has been established that there has been a loss of this nature. To do this the contractor must be able to show that, but for the delay, the contractor would have been able to earn the amounts claimed on another contract, eg by producing evidence such as invitations to tender which were declined. Such formulae may be useful where it is difficult to quantify the amount of the alleged loss, provided a check is made that the assumptions on which the formula is based apply.

6.27   Although direct loss and/or expense is a matter of money not time, which are quite separate issues, there is often a practical correlation in the case of prolongation. Any general implication, however, that there is a link would be incorrect and in principle disruption claims and delay to progress are independent. An extension of time, for example, is not a condition precedent to the award of direct loss and/or expense *(H Fairweather & Co v Wandsworth)* *(see para 4.28 above)*.

6.28   Though the architect may delegate the duty of ascertaining the direct loss and expense, it does not appear that it is obligatory to accept the quantity surveyor's opinion *(R Burden v Swansea Corporation)*, although the quantity surveyor's assessment would be strong evidence as to what the correct amount should be. Applications or claims from the contractor made under the contract must be dealt with according to the procedures of the contract. Failure to certify an amount properly due will not prevent recovery, and could leave the employer liable in damages for breach of contract *(Croudace v London Borough of Lambeth)*.

R Burden Ltd v Swansea Corporation [1957] 3 All ER 243 (HL)

Burden entered into a contract with Swansea Corporation to build a school. The contract provided for interim certificates to be issued at intervals by the architect. The architect, who was the council's Borough Architect, acted originally as both architect and surveyor under the contract. Later, after 20 certificates had been issued, the firm of quantity surveyors who had originally prepared the Bills were appointed to act as surveyors under the contract in place of the Borough Architect. In the next certificate the surveyor reduced the amount applied for by the contractor by around 75 per cent, and the architect certified the lower figure. The surveyor later discovered that he had made a mistake, but did not inform the architect of the error. The contractor gave notice determining the contract, on the grounds that the employer had interfered with the issue of the certificate. The House of Lords decided that a mistake in a direction as to the amount to be paid did not amount to interference or obstruction. It was suggested that the architect would have been at liberty to have certified a different amount if aware of the error.

Croudace Ltd v London Borough of Lambeth (1986) 33 BLR 25 (CA)
Croudace entered into an agreement with the London Borough of Lambeth to erect 148 dwelling houses, some shops and a hall. The contract was on JCT63 and the architect was the council's Chief Architect and the quantity surveyor was the council's Chief Quantity Surveyor. The architect delegated his duties to a private firm of architects. Croudace alleged that there had been delays and that they had suffered direct loss and/or expense and sent letters detailing the matters to the architects. In reply, the architects told Croudace that they had been instructed by Lambeth that all payments relating to 'loss and expense' had to be approved by the council. The Chief Architect then retired and was not immediately replaced. There were considerable delays pending a further appointment and Croudace began legal proceedings. The High Court found that the council was in breach of contract in failing to take the necessary steps to ensure that the claim was dealt with, and were liable to Croudace for this breach. The Court of Appeal upheld this finding.

Fluctuations

6.29   In some projects it may be an advantage to insist on literally a 'fixed' or 'guaranteed' price, whereby the contractor accepts the risk of all changes in the cost of the Works due to statutory revisions and market price fluctuations. However, requiring this degree of certainty will, in some economic climates, result in higher tender figures as the contractor will need to allow for possible increases, particularly if the contract period is relatively lengthy. In order to avoid the inflated tenders, most contracts allow for some 'fluctuations', whereby the employer accepts some of these risks.

6.30   In IFC98 the fluctuations provisions are covered in clause 4.9 and 4.10, which refer to Supplemental Conditions C and D (published separately). Clause 4.9 covers fluctuations to the adjusted contract sum less amounts included for named sub-contractors, and clause 4.10 covers fluctuations in respect of the work of named sub-contractors. Supplemental Condition C is for 'Tax etc' fluctuations; and Supplemental Condition D covers the use of price adjustment formulae. Clause 4.9 states that unless Supplemental Condition D is identified in the Appendix, and contract Bills are included in the contract documents, Supplemental Condition C applies.

6.31   Supplemental Condition C provides for full recovery of all fluctuations in the rates of contributions, levies and taxes in the employment of labour, and in the rates of duties and taxes on the procurement of materials. In short, the only amounts payable are those arising out of an Act of Parliament or delegated legislation. Contractors, however, have pointed out that many less obvious increases are nevertheless not included, therefore a 'percentage addition' is made to allow for these. The agreed percentage is entered in the Appendix. Supplemental Condition D allows for adjustment based on the use of formulae: it does not necessarily take account of the actual costs, but is relatively simple to operate, and is generally considered by contractors to be a fair adjustment.

6.32   Where a contract includes for fluctuations, in the absence of anything to the contrary, they will be payable for the whole time the contractor is on site even though it fails to complete within the contract period *(Peak Construction (Liverpool) Ltd v McKinney Foundations Ltd)*. There is a so-called 'freezing' provision in IFC98 clauses C4.7 and D12.1, but this depends on the clause 2.3, 2.4 and 2.5 text being left unamended, and all notices of delay properly dealt with by the architect.

Peak Construction (Liverpool) Ltd v McKinney Foundations Ltd (1970) 1 BLR 111 (CA)
Peak Construction were main contractors on a contract to construct a multi-storey block of flats for Liverpool Corporation. As a result of defective work by nominated sub-contractors McKinney Foundations' work on the main contract was halted for 58 weeks, and the main contractors brought a claim against the sub-contractors for damages. The Official Referee at first instance found that the entire 58 weeks delay was caused by the nominated sub-contractor, and awarded £40,000 damages, £10,000 of which was for rises in wage rates during the period. McKinney appealed, and the Court of Appeal found that the award of £10,000 could not be upheld as clause 27 of the main contract entitled Peak Construction to claim this from the council right up until the time when the work was halted.

6.33   The fluctuations to be added to the contract sum in respect of named sub-contractors are the amounts resulting from the application of the fluctuations provisions in clauses 33 and 34 of NAM/SC. These are also frozen once the sub-contractor is in a period of culpable delay. In addition, should the period in which the sub-contractor is to complete its work be extended on account of a default of the main contractor, the net amount of fluctuations resulting from the period of extension, although added to the monies due to the sub-contractor, is not added to the contract sum. In other words although the main contractor must pay to the sub-contractor fluctuations for a period of delay which it has caused the sub-contractor, this amount is not reimbursed by the employer.

7.1   One of the most important duties of the architect under IFC98 is the issuing of certificates of payment. Failure to carry out this duty with a reasonable degree of care will put the employer at considerable risk. It is not unheard of for contractors to become insolvent during the course of a contract, and if the certificates have been overvalued the employer may suffer losses which could have been avoided.

7.2   On the other hand, the contractor has a right to be paid what the terms of the contract state is due, and the provisions introduced through the Housing Grants, Construction and Regeneration Act 1996 were partly as a result of the common occurrence of employers withholding payment without due cause. The architect must therefore administer the provisions fairly, and must not be influenced by any attempt on the part of the employer to delay certification or reduce the amounts shown to suit its own economic circumstances.

7.3   It is quite clear from the well known case of *Sutcliffe v Thackrah* that failure to certify correctly could amount to a breach of duty to the employer. The position is less clear with respect to the architect's duty of care to the contractor. A certifier was found liable to a contractor in the case of *Michael Salliss & Co v Calil and W F Newman & Associates*. Although this appeared to be overtaken in *Pacific Associates Inc v Baxter* the latter involved non-standard clauses which, if they had not been present, may have resulted in a different outcome.

Sutcliffe v Thackrah (1974) 4 BLR 16 (CA)
An architect issued certificates on a contract for the construction of a dwelling house. The contractor's employment was determined for proper reasons following which the contractor went bankrupt. It then became apparent that much of the work, which had been included in the interim certificates, was defective, and the architect was found negligent. In the House of Lords, when reviewing the role of the architect, Lord Reid stated: 'Many matters may arise in the course of the execution of a building contract where a decision has to be made which will affect the amount of money which the contractor gets ... the building owner and the contractor make their contract on the understanding that in all such matters the architect will act in a fair and unbiased manner and it must therefore be implied into the owner's contract with the architect that he shall not only exercise due skill and care but also reach such decisions fairly holding the balance between his client and the contractor' (at page 21).

Michael Salliss & Co Ltd v Calil and William F Newman & Associates (1987) 13 ConLR 69
Calil employed contractor Michel Salliss for some refurbishment works on JCT63. W F
Newman acted as architect and quantity surveyor under the contract. The contractor
commenced proceedings against the employer and joined the architects as second
defendants, claiming that the architect was in breach of his duty to use all professional
skill and care in granting only a 12-week extension of time when a 29-week extension
was due. There was a sub-trial as to whether the contractor could recover damages
against the architect. His Honour Judge Fox-Andrews held that under a JCT contract the
architect owed a duty to the contractor to act fairly between the employer and contractor
in matters such as certification and extensions of time. He also noted that: 'in many
respects an architect in circumstances such as these owes no duty to the contractors.
He owes no duty to contractors in respect of the preparation of plans and specifications
or in deciding matters such as whether or not he should cause a survey to be carried
out. He owes no duty of care to a contractor whether or not he should order a variation.
Once, however, he has ordered a variation he has to act fairly in pricing it' (at page 79).

Pacific Associates Inc v Baxter (1987) 13 ConLR 80
Pacific Associates contracted on the FIDIC form of contract to carry out dredging and
reclamation works for the Ruler of Dubai. The defendant was employed as engineer to
administer the contract. Disputes arose between the employer and contractor which
went to arbitration, and were subsequently settled with a £10 million payment to the
contractor, with both parties paying their own costs. The contractor then brought a claim
against the engineers for negligent certification, claiming the unrecovered balance of the
claim together with interest and arbitration costs. His Honour Judge John Davies
dismissed the claim. The contract contained a particular condition which stated: 'Neither
any member of the Employer's staff nor the Engineer nor any of his staff, nor the
Engineer's Representative shall be in any way personally liable for the acts or obligations
under the Contract, or answerable for any default or omission on the part of the Employer
in the observance or performance of any of the acts, matters or things which are herein
contained'. The judge stated that 'the clear intention of (this clause) ... was to relieve the
engineer of all personal liability for his acts and obligations under the contract' (at page
93). He also stated that he felt the question of liability always depended on the particular
terms of the contract in question, and that 'the over-riding intention of the contract was
to put the engineer beyond the reach of legal responsibility for his acts' (at page 92). The
contractor appealed but the appeal was dismissed. The Court of Appeal stated that the
existence of the special clause meant that a duty of care could not in this case be
imposed, but emphasised that otherwise such a duty might have existed.

7.4   The payment provisions in IFC98 comply with the Housing Grants, Construction and Regeneration Act 1996, and also follow Latham's recommendations. The result is a quite complex set of payment provisions, perhaps in some respects out of proportion with what had previously been a relatively simple form. Nevertheless the provisions of the 1996 Act are of course compulsory for all contracts to which the Act applies, and the remaining clauses should not be amended without taking expert advice.

7.5   Interim payments are to be made by the employer to the contractor after the issue of certificates by the architect (cl 4.2(a)). Unless there has been an agreement between the parties that payments will be in relation to work stages, certification will normally be at monthly intervals up until the date of Practical Completion. The Appendix allows the parties to enter a date for the first certificate; if no date is entered the first certificate must be issued within one month of the date of possession. Certificates are issued to the employer and a duplicate sent to the contractor (cl 1.9).

7.6   There is also an optional provision for advance payment to the main contractor, provided the employer is not a local authority. An entry must be made in the Appendix to show whether or not the optional provision is to apply. If it is, the amount will be entered as either a fixed sum, or a percentage of the contract sum. The entry must also show when it is to be paid to the contractor, and when it is to be reimbursed to the employer. A bond may be required, and IFC98 includes an advance payment bond as Annex 1 to the Appendix.

7.7   There is always a risk in making an advance payment with respect to a construction contract, even when backed by a bond, and the procedure will inevitably involve extra expense to the employer. The employer should be quite clear as to what compensatory benefits, such as a reduction in the contract sum, will result before agreeing to any arrangement of this sort. In the end it is the employer's decision, but the architect may need to explain the provisions and give initial advice.

Valuations and ascertainment of amounts due

7.8   There are two methods by which the amount due to the contractor as interim payments can be assessed: either through acceptance of the figure stated in a contractor's application for payment, or through valuation by the quantity surveyor. The latter has traditionally been the basis for assessing payments under JCT forms.

7.9   Clause 4.2(c) gives the contractor the right to submit its own assessment of the value of an interim certificate. The application must be made at least seven days before the certificate is due and should be sent direct to the quantity surveyor. It is then open to the quantity surveyor to agree or disagree with this valuation; if the quantity surveyor disagrees with the amount he or she must identify in detail the matters where the differences arise. The architect should request that the quantity surveyor forwards copies of all correspondence and keeps the architect informed regarding applications.

7.10   If no application is received, or if the application is rejected, the valuation is carried out by the quantity surveyor as directed by the architect and should cover the amounts due at a date not more than seven days before the issue of the certificate. Valuations can be made 'whenever the architect considers them necessary' except when the fluctuation formula adjustment procedure is used, where timing cannot be at the discretion of the architect (cl 4.2(c)). The roles of the quantity surveyor and the architect, however, are quite distinct, and the valuation figure will not necessarily be the same as that on the interim certificate.

Coverage of the certificate

7.11   Clause 4.2(a) requires that interim certificates state not only the amount to be paid, but also 'to what the amount relates and the basis on which that amount was calculated'. It is unlikely that a great deal of detail will be required here, a short schedule will probably be sufficient. Similar provisions are included for the Final Certificate.

7.12   The amount certified is the total amount ascertained under clause 4.2.1 and 4.2.2, less any amounts of advance payment due for reimbursement. The amounts can be summarised as follows.:
95 per cent of the following (ie items subject to retention) (cl 4.2.1):
- total value of work properly executed (cl 4.2.1(a));
- total value of materials and goods properly on site (cl 4.2.1(b));
- value of off-site materials, goods or items, provided they are 'listed items' (cl 4.2.1(c));
100 per cent of the following items, if applicable (cl 4.2.2):
- additional insurance premium related to employer's occupation of the site (cl 2.1);
- costs of opening up and tests (cl 3.12);
- fluctuations (cl 4.9 and 4.10);
- direct loss and/or expense (cl 4.11);
- statutory fees and charges (cl 5.1);
- insurance monies (cl 6.2.4 and cl 6.3);

- sums for restoration of damaged work (cl 6.3).

Value of work properly executed

7.13 The architect should only certify after having carried out an inspection to a reasonably diligent standard. Architects should not include any work that appears not to have been properly executed, whether or not it is about to be remedied or the retention is adequate to cover remedial work *(Townsend v Stone, Sutcliffe v Chippendale & Edmondson)*. Where work, which has been included in a certificate, subsequently proves to be defective the value can be omitted from the next certificate, even to the extent that it produces a negative amount, ie a repayment by the contractor.

7.14 The value of the work will be calculated using the rates shown in the priced document, whichever is appropriate. If a priced activity schedule is included, the amount shown on any interim certificate in respect of any items listed in the activity schedule should represent 95 per cent of the total of the amounts reached by multiplying the percentage of the work properly executed by the price for that work as shown on the activity schedule. The fact that an activity schedule is used, however, does not lessen the architect's duty to determine that all work certified has been carried out in accordance with the contract.

Townsend v Stone Toms & Partners (1984) 27 BLR 26 (CA)

Mr Townsend engaged architects Stone Toms in connection with the renovation of a farmhouse in Somerset. John Laing Construction Ltd were employed to carry out the work on JCT67 Fixed Fee Form of Prime Cost Contract. Following the end of the defects liability period the architects issued an interim certificate that included the value of work which they had already included in their schedule of defects, and which they knew had not yet been put right. Mr Townsend brought proceedings against both Stone Toms and Laing. Laing made a payment into court of £30,000, which was accepted by Townsend in full and final settlement. Townsend then continued with the proceedings against the architect, claiming that he was entitled to recover any excess that he might have obtained for Laing had he continued with those proceedings. The Official Referee assessed the total value of the claims against Laing as only £25,000, therefore no excess was recoverable. The Deputy Official Referee also found that the architect was not negligent in issuing the interim certificate. Mr Townsend appealed and the Court of Appeal, although approving the lower court's decision on the effect of the payment into court, held that the architect had been negligent. Oliver LJ stated 'the whole purpose of the certification is to protect the client from paying to the builder more than the proper value of the work done, less proper retention, before it is due. If the architect deliberately over-certifies work which he knows has not been done properly, this seems to be a clear breach of his contractual duty, and whether certification is described as "negligent" or "deliberate" is immaterial' (at page 46).

Sutcliffe v Chippendale & Edmondson (1971) 18 BLR 149
(Note this case is the first instance decision which was appealed to the Court of Appeal *sub nom Sutcliffe v Thackrah*, discussed above.)
Mr Sutcliffe engaged the architects Chippendale & Edmondson in relation to a project to build a new house. No terms of engagement were agreed, but the architects proceeded to design the house, invite tenders, and arrange for the appointment of a contractor on JCT63. Work progressed slowly and towards the end of the work it became obvious that much of the work was defective. The architect had issued ten interim certificates before Mr Sutcliffe entirely lost confidence, dismissed the architect and threw the contractor off the site. He then had the work completed by another contractor and other consultants which cost around £7,000, in addition to which he was obliged, as a result of the original contractor having obtained judgment against him, to pay all ten certificates in full. As this contractor then went bankrupt he then brought a claim against the architects. The architects contended, among other things, that their duty of supervision did not extend to informing the quantity surveyor of defective work that should be excluded from the valuation. His Honour Judge Stabb QC found for Mr Sutcliffe, stating 'I do not accept that the words "work properly executed" can include work not then properly executed but which it is expected, however confidently, the contractor will remedy in due course' (at page 166).

Unfixed and off-site materials and goods

7.15   The interim certificate should include materials which have been delivered to the site but are not yet incorporated in the Works (cl 4.2.1(b)). In spite of detailed provisions aimed at protecting the employer, there remains some risk in including these items. Once materials have been built in, under common law they would normally become the property of the owner of the land, irrespective of whether or not they have been paid for by the contractor. This would be the case even if there were a retention of title clause in the contract with the sub-contractor or supplier. A retention of title clause is one which stipulates that the goods sold do not become the property of the purchaser until they have been paid for, even if they are in the possession of the purchaser.

7.16   The employer could be at risk, however, where materials have not yet been built in, even where the materials have been certified and paid for. The contractor might not actually own the materials paid for because of a retention of title clause in the sale of materials contract. Under the Sale of Goods Acts 1979 ss 16–19, property in goods normally passes when the purchaser has possession of them, but a retention of title clause will be effective between a supplier and a contractor even where the contractor has been paid for the goods, provided they have not yet been built in. It should be noted, however, that the employer may have some protection through the operation of s 25 of the Act, which in some circumstances allows the employer to treat the contractor as having authority to transfer the title in the goods, even though this may not in fact be the case *(see Archivent v Strathclyde Regional Council below)*.

7.17 Another risk relating to rightful ownership is where the contractor fails to pay a domestic sub-contractor who has purchased materials, and the sub-contractor claims ownership of the unfixed materials. Here the risk may be higher, as a work and materials contract is not governed by the Sale of Goods Act. Therefore there can be no assumption that property would pass on possession.

7.18 IFC98 attempts to deal with the issues surrounding ownership in several ways. First, unfixed materials and goods, which have been delivered to the site and intended for the works, may not be removed without the written consent of the architect (cl 1.10). Removal would be a breach of contract, therefore the employer could claim from the contractor for any losses suffered through unauthorised removal. This would apply even though the materials or goods may not yet have been included in any certificate. Secondly, unfixed materials and goods either on or off site which have been included in a certificate which has been paid, are to become the property of the employer (cl 1.10), and the contractor is thereby prevented from disputing ownership.

7.19 Clause 1.10 of the main contract, however, is only binding between the parties, and does not place obligations on any sub-contractor. The risk facing the employer is that if the contractor becomes insolvent a sub-contractor or supplier may still have a rightful claim to ownership of the unfixed goods, even though they have been paid for by the employer *(see Dawber Williams Roofing v Humberside County Council)*. The main contract therefore requires that all sub-contracts include similar clauses to 1.10 regarding non-removal from site, and ownership passing upon payment (cl 3.2.2(a) and (c)). Sub-contracts must also include a clause stating that once materials and goods have been certified and paid for under the main contract they become the property of the employer and that the sub-contractor 'shall not deny' this (cl 3.2.2(b)). This would operate even where the main contractor has become insolvent. Even this might not protect the employer in some circumstances, because if the sub-contractor does not have 'good title' it cannot pass it on. Thus, for example, it might not prevent a sub-sub-contractor claiming ownership.

Archivent Sales & Developments Ltd v Strathclyde Regional Council (1984) 27 BLR 98
(Court of Session, Outer House)
Archivent agreed to sell a number of ventilators to a contractor who was building a primary school for Strathclyde Regional Council. The contract of sale included the term 'Until payment of the price in full is received by the company, the property in the goods supplied by the company shall not pass to the customer'. The ventilators were delivered and included in a certificate issued under the main contract (JCT63), which was paid. The contractor went into receivership before paying Archivent, who claimed against the council for the return of the ventilators or a sum representing their value. The council claimed that s 25(1) of the Sale of Goods Act 1979 operated to give them an unimpeachable title. The judge found for the council. Even though the clause in the sub-contract successfully retained the title for the sub-contractor, the employer was entitled to the benefit of s 25(1) of the Sale of Goods Act. The contractor was in possession of the ventilators and had ostensible authority to pass the title on to the employer, who had purchased them in good faith.

Dawber Williams Roofing Ltd v Humberside County Council (1979) 14 BLR 70
The plaintiffs entered into a sub-contract with Taylor and Coulbeck Ltd (T&C) to supply and fix roofing slates. The main contractor's contract with the defendant was on JCT63. By clause 1 of their sub-contract (which was on DOM/1) the plaintiffs were deemed to have notice of all the provisions of the main contract, but it contained no other provisions as to when property was to pass. The plaintiffs delivered 16 tons of roofing slates to the site, which were included in an interim certificate, which was paid by the defendant. T&C then went into liquidation without paying the sub-contractor, who brought a claim for the amount or alternatively the return of the slates. The judge allowed the claim, holding that clause 14 of JCT63 could only transfer property where the main contractor had a good title. (The difference between this and the *Archivent* case above is that in this case the sub-contract was a contract for work and materials, to which the Sale of Goods Act 1979 did not apply.) Provisions within clause 19.4 of IFC98 now deal with the problem illustrated by this case.

7.20   Under the IFC98 provisions the architect is obliged to include the unfixed materials in an interim certificate, even though this limited risk to the employer remains, provided that the materials are not prematurely delivered, or not properly protected. Architects should pay careful attention to the exact wording of this Condition.

'Listed items'

7.21   Interim certificates might include amounts in respect of any 'listed items' (cl 4.2.1(c)). If this provision is to apply then the list must be attached to the Bills of Quantities (or specification/schedules of work) to which the contractor has tendered. The listed items may be 'uniquely identified' or not uniquely identified (ie materials or goods or items prefabricated for inclusion in the Works). The value of items listed must be included in an interim certificate prior to delivery on site, provided certain pre-conditions are fulfilled:

- the contractor has provided reasonable proof that the property is vested in it (cl 4.2.1(c).1);
- if the item is not 'uniquely identified', or if required in the Appendix the contractor has provided a bond (cl 4.2.1(c).1 and .2);
- the listed items are in accordance with the contract (cl 4.2.1(c).3);
- the listed items are 'set apart' or clearly marked and identified (cl 4.2.1(c).4);
- the contractor provides proof that the items are insured against Specified Perils until delivery on site (cl 4.2.1(c).5).

7.22   The architect has no discretionary power to certify any off-site items, other than those that have been listed. This makes the position for both parties clear, in that only 'listed' off-site materials are to be certified. The architect should therefore be careful not to include any unlisted off-site materials in any certificate, as only listed items are covered by the clause 1.11 safeguards to the employer regarding ownership and responsibility for loss or damage.

Other items in the gross valuation

7.23   In addition to the value of work properly executed and of materials properly on the site, clause 4.2.2 lists out other amounts that must be included in the gross valuation for an interim certificate. For example, payments ascertained as due to the main contractor under clause 4.9(a) (tax, etc fluctuations), or under clause 4.11 (direct loss and/or expense) are to be added to the contract sum.

Deductions from the gross valuation
Withheld percentage

7.24   Some of the items that must be included in the gross valuation are subject to a reduction of 5 per cent (cl 4.2.1). Half of the withheld amount is released upon Practical Completion, and the remaining half with the final certificate. The amount is commonly referred to as 'retention' although IFC98 does not use this term. There is no opportunity to enter an alternative percentage – if another figure is required the form would have to be altered. The employer is trustee of the withheld percentage for the contractor (cl 4.4), but is not obliged to invest it for the contractor, and would have the benefit of any interest which accrues. The employer has the right to deduct from the retention sums due from the contractor, including sums due by right of set-off.

7.25   Although IFC98 contains no provision requiring the employer to place the withheld percentage in a separate banking account, a series of cases have established that such an obligation would be implied in any contract that requires the employer to hold the money as a trustee (see Rayack v Lampeter, Wates Construction v Franthom Property, and Finnegan v Ford Sellar Morris). However, the court will not make an order to place money in a separate account following the insolvency of the employer (see Mac-Jordan Construction v Brookmount Erostin). The contractor would have no special claim beyond that of an unsecured creditor. To be safe the contractor must insist, while the employer is solvent, that the money is placed in a separate account. This dilemma over retention and the effectiveness of trustee status has raised the question of bonds and guarantee bonds from both employer and contractor, respectively, as an alternative.

Rayack Construction Ltd v Lampeter Meat Co Ltd (1979) 12 BLR 30
Rayack Construction agreed to build two meat-processing plants for Lampeter Meat Company at their factory. The contract was a modified version of JCT63 that required payment of certified amounts within 90 days of the date on the certificate, and allowed for 50 per cent retention to be deducted by the employer. The contract also contained a clause similar to clause 30.5.1, which stated that the employer's interest in the retention was fiduciary as trustee for the contractor, but contained no provision requesting that the retention be held in a separate bank account. The court held, nevertheless, that the employer was under an obligation to place the fund in a separate account. The clause would otherwise be of no practical application as it would not protect the contractor from the consequences of the employer's insolvency.

Wates Construction (London) Ltd v Franthom Property Ltd (1991) 53 BLR 23 (CA)
Wates entered into a contract with Franthom on JCT80 to construct a hotel in Kent. Clause 30.5.3 (requiring Franthorm to place retention in a separate account) had been deleted, but otherwise the retention clauses were in all material respects the same as those in IFC98. Although requested to by Wates, Franthom refused to place the accrued retention of around £84,000 in a separate account. Wates then commenced legal

proceedings. Judge Newey ordered Franthorm to place the money in an account, and Franthom then appealed. The court dismissed the appeal stating that 'clear express provisions are needed if a separate bank account is not to be set up'. The fact that the clause had been deleted did not of itself indicate what the parties intentions were, the effect was the same as if the words had never been there at all.

J F Finnegan Ltd v Ford Sellar Morris Developments Ltd (1991) 53 BLR 38
Finnegan were contractors on a JCT81 contract for works at Ashford. After the works reached practical completion the employer claimed liquidated damages of around £60,000 against a sum admitted as due to the contractor of around £20,000. Under clause 30.4.2.2 the employer was obliged to place retention monies deducted in a separate account if requested by the contractor. Finnegan commenced action to recover the sum due. The employer counterclaimed for the liquidated damages and Finnegan then requested that the retention be placed in a separate account. The employer refused and Finnegan applied for an injunction. The judge granted the injunction, despite the fact that this was long after practical completion. The contract did not require that a request was made each time retention was deducted nor at the time it was deducted.

Mac-Jordan Construction Ltd v Brookmount Erostin Ltd (1991) 56 BLR 1
A developer held over £100,000 for the contractor in retention money but was also heavily indebted to the bank (floating loan granted by a charge). The developer went into insolvency and the bank appointed administrative receivers. The contractor then sought a court injunction to establish a separate retention fund but the Court of Appeal refused on grounds that this would give an unsecured creditor (the contractor) preference over any other unsecured creditors of an insolvent debtor. The contractor's right to the retention was stated to be no more than an 'unsatisfied and unsecured contractual right for the payment of money' (Scott LJ at page 15).

Advance payments and bonds

7.26 If the parties have elected to use the advance payment provisions, the payment indicated in the Appendix is paid to the contractor before the first certificate of payment is due for issue, but only after the contractor has provided the bond required **(cl 4.2(b))**. Payment is made direct from the employer to the contractor, and the architect should ensure receipt of copies of any correspondence regarding this. Details of when the reimbursements are to take place will also be set out in the Appendix and could, eg, be in stages throughout the project. The reimbursement is deducted from the gross valuation under the relevant certificate. It should be noted that the amount to be deducted each month should be a cumulative total of the reimbursements, with the final deduction equalling the original advance payment, and that final deduction appearing on all subsequent certificates including the Final Certificate. The current wording of the contract does not make this entirely clear, and it is understood that the JCT are considering making some minor adjustments.

VAT

7.27 The contract sum is exclusive of any VAT (cl 5.5). There are supplemental provisions to the contract (the VAT Agreement) and an Appendix entry should indicate whether clause 1A applies. This will affect the procedure for VAT payments.

Payment procedure

7.28 The final date for payment of each interim certificate is 14 days from the date of issue (cl 4.2(a)). It should be noted that 14 days commences with the date on the certificate, not the date of its receipt by the employer, and that the payment should reach the contractor by the final date for payment, so that allowance should be made for posting.

7.29 Clause 4.2.3 sets out requirements for giving notice with regard to interim certificates. These provisions are required by the Housing Grants, Construction and Regeneration Act 1996, ss 110 and 111, and are repeated with respect to the Final Certificate. The employer must give the contractor notice of how much it intends to pay no later than five days after an interim certificate is issued (cl 4.2.3(b)). The clause follows the wording of the Act in that the notice is required whether or not the employer intends to make any deduction (see figure 9). In practice there would be no significant loss to the contractor if the employer failed to issue this notice when no deduction was intended *(see Appendix B)*, but the intention of the provision was that the contractor should be fully aware of the payment it is to receive, and without confirmation there may remain a degree of uncertainty.

7.30 The contract states that the employer 'may' give written notice of its intention to withhold any amount from the due amount no later than five days before the final date for payment, clearly stating the grounds for making the deduction (cl 4.2.3(c)). The contract then states 'Where the Employer does not give any written notice pursuant to clause 4.2.3(b) and/or clause 4.2.3(c) the Employer shall pay the Contractor the amount due pursuant to clauses 4.2.1 and 4.2.2' (cl 4.2.3(d)). The effect of these two clauses is that if the employer intends to withhold any amount then it *must* issue a notice to that effect under the

Figure 9 Payment procedure

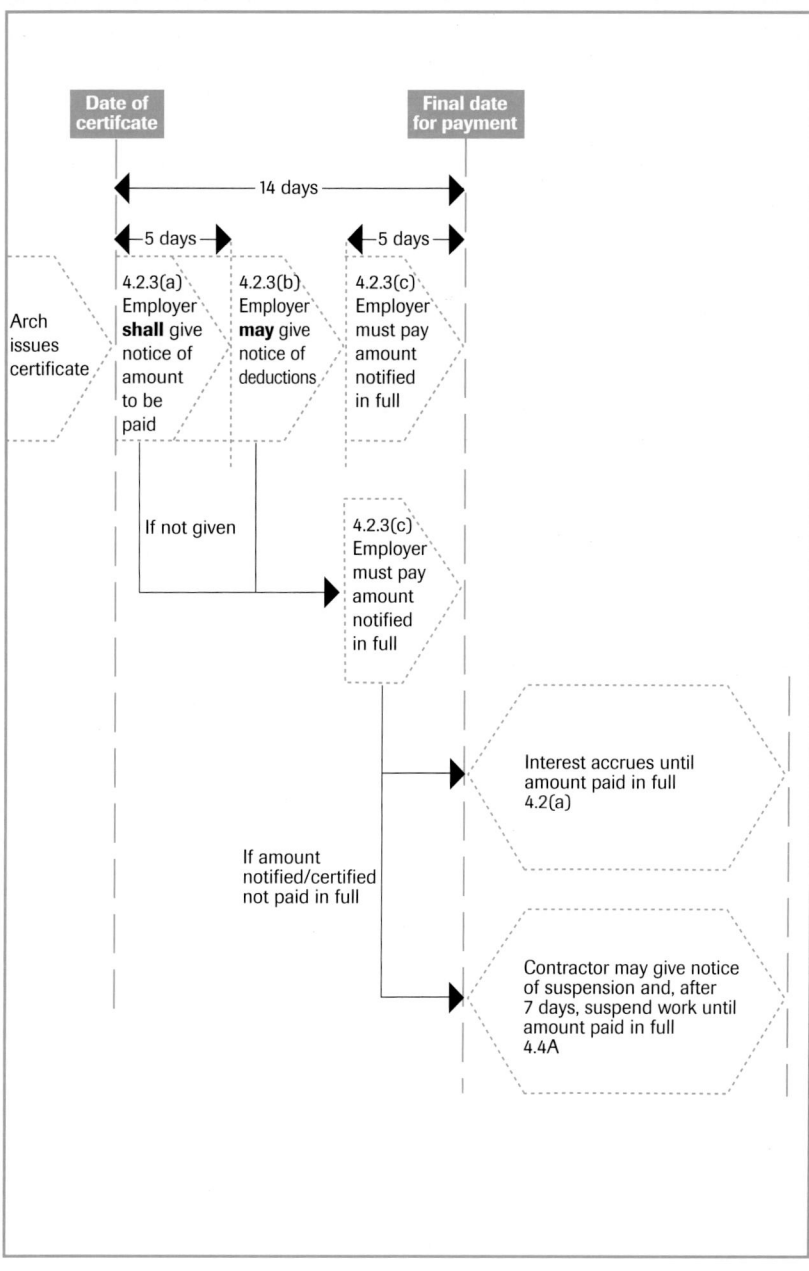

contract. It is suggested that if it is known in advance that a deduction is intended then the first notice under clause 4.2.3(c) *must* make this clear. The second notice may then be superfluous, always provided that the first notice included all the necessary information required under clause 4.2.3(b). If not all information is given, or new circumstances develop following the first notice stage, then the second notice will be required. In any event many employers might prefer to adopt the cautious and protective line of issuing both notices whenever a deduction is intended.

Deductions

7.31   The contract expressly gives the employer the right to make certain deductions from the certified sums due to the contractor. Note that these are distinct from amounts that may be deducted before arriving at the sum to be shown on the certificate, for example deductions from the contract sum where incorrect setting out has been accepted (cl **3.9**). Notice of the deduction should have been given as described above. Deductions authorised by the contract which the employer may make before payment of certified amounts arise in respect of:
- losses due to non-compliance with architect's instructions (cl **3.5.1**);
- default in taking out or maintaining insurance premium (cl **6.2.3 and 6.3A.2**);
- payment or allowance of liquidated damages (cl **4.7.1**);
- the making of statutory tax deductions (Supplemental Condition B).

The Construction Industry Scheme

7.32   The Construction Industry Scheme (CIS), which took effect on 1 August 1999, has replaced the statutory tax deduction scheme which was previously in force, and the provisions in IFC98 relating to statutory tax were consequently revised through Amendment 1 to the form. Like the previous scheme it is founded in the Income and Corporation Taxes Act 1988, and is intended largely as a measure to combat tax evasion, particularly where it occurred in connection with 'lump' labour. Under certain circumstances the employer will operate on behalf of the Inland Revenue and collect tax due.

7.33   Supplemental Condition B takes effect where the employer is a 'contractor' for the purposes of the 1988 Act. The terminology of the Act is confusing because the employer could be a 'contractor' and the building contractor a 'sub-contractor'. A 'contractor' has a broad definition under the Act and includes anyone who regularly carries out or commissions construction work on their own premises or investment properties, and whose average annual turnover exceeds £1 million.

7.34   If the employer is a 'contractor' then the employer must not make any payment to the building contractor unless the contractor has provided a valid authorisation **(cl B3)**. The definition of a valid 'Authorisation' is given in clause B1 and it could be either a CIS registration card or certificate issued by the Inland Revenue, or a 'certifying document' in the form prescribed by the Income Tax (Sub-contractors in the Construction Industry) Regulations 1993 (SI 1993/743).

7.35   If there is a valid certificate or certifying document the employer may pay the contractor the full amount certified. If the contractor supplies the CIS registration card the employer must take off from the payment the required 'statutory deduction'. If no valid authorisation is supplied the employer must notify the contractor under clause B4.2 and may not make any payment until the authorisation is received.

7.36   If the employer is not a 'contractor' and does not become one during the contract, the provisions in Supplemental Condition B will not apply. The architect should remind the employer of its statutory obligations and draw attention to the provisions in the contract at an early stage in the project. However, the legislation places the onus on the employer, if a 'contractor', to make sure that the CIS is correctly applied. The provisions are explained briefly in JCT Practice Note 1 (series 2) *(see Appendix C)*.

Contractor's position if the certificate is not paid

7.37   IFC98 includes several provisions which protect the contractor if the employer fails to pay the contractor amounts due. Clause 4.2(a) makes provision for simple interest on late payments of certificates. This is set at 5 per cent over the Base Rate of the Bank of England, and the interest accrues from the final date for payment until the amount is paid. Similar provisions are included for the Final Certificate. (It should be noted that if the provision were deleted the contractor would normally have a statutory right to interest under the Late Payment of Commercial Debts (Interest) Act 1998.) If the employer makes a valid deduction following a notice it is suggested that interest would not be due on this amount. The clause does not refer to the amount stated on the certificate but to 'the amount due to the Contractor under the Conditions' which would take into account valid deductions.

7.38   The contractor is also given a 'right of suspension' under clause 4.4A. This right is required by the Housing Grants, Construction and Regeneration Act 1996. If the employer fails to pay the contractor by the final date for payment the contractor has a right to suspend performance of all its obligations under the contract, which would include not only the carrying out of the work but, eg, could also extend to any insurance obligations. This right is stated to be 'subject to any notice issued pursuant to clause 4.2.3(b) or 4.3(c)', which suggests that the contractor may not suspend work if a notice to withhold payment has been given by the employer. The contractor must have given the employer written notice of its intention to suspend work and stated the grounds for the suspension, and the default must have continued for a further seven days. The contractor must resume work when the payment is made. Under these circumstances the suspension would not give the employer the right to determine the contractor's employment. Any delay caused by the suspension could be a Relevant Event (cl 2.4.18) and a 'matter' in relation to direct loss and/or expense (cl 4.12.10).

7.39   The contractor also has the right to determine the contract if the employer does not pay amounts properly due (cl 7.9.1(a)). The contractor must give notice of this intention, which specifies the default as required by the contract *(see para 9.25 below)*.

Contractor's position if it disagrees with amount certified

7.40   The issue of a certificate is a condition precedent of the right of the contractor to be paid *(Lubenham v South Pembrokeshire District Council)*. This case states that the contractor is only entitled to the sum stated in the certificate, even if the certificate contains an error, for example because it includes a wrongful deduction. The contractor should note that it has no right to suspend work, or to determine the contract, simply because it feels that certificates have been undervalued (the author has come across instances of both of these in recent months). The contractor's remedy is to request that the error is corrected in the next certificate, or to bring proceedings to have the certificate adjusted. There are exceptions to this rule, where, for example, the employer has interfered with the issue of the certificate, in which case the contractor may be entitled to summary judgment for the correct amount.

Lubenham Fidelities and Investments Co Ltd v South Pembrokeshire District Council (1986) 33 BLR 39 (CA)

Lubenham Fidelities was a bondsman who elected to complete two building contracts, both based on JCT63. The architects, Wigley Fox Partnership, issued several interim certificates which stated the total value of work carried out, but also made deductions for liquidated damages and defective work from the face of the certificate. Lubenham protested that the certificates had not been correctly calculated, withdrew their contractors from the site and issued notices to determine the contract. Shortly after, the council gave notice of determination of the contract. Lubenham brought a claim against the council claiming that their notices were valid and effective, and against Wigley Fox on the basis that their negligence had caused them losses. It was held that the council were not obliged to pay more than the amount on the certificate, and that whatever the cause of the undervaluation the correct procedure was not to withdraw labour, but to request that the error was corrected in the next certificate, or to pursue the matter in arbitration. Lubenham's claim against Wigley Fox failed because it had been the suspension of the works rather than the certificates that had caused the losses, and because the architects had not acted with the intention of interfering with the performance of the contract.

Employer's obligation to pay

7.41   As discussed at para 7.31 above, the contract authorises the employer to make certain deductions from amounts certified, and if this is intended then the required notices must be issued. The contract does not discuss, however, the position where the employer disagrees with the amount shown on the certificate. This may occur, for example, where the employer feels that work has been included which is defective, or where the architect has failed to make a deduction for incorrect setting out under clause 3.9. The employer could issue a notice declaring an intention to pay less than the certified amount, and raise the matter using one of the dispute resolution methods provided for in the contract, and request the tribunal to use its powers to revise the certificate. Alternatively the employer could issue the notice and allow the contractor to raise the dispute if it so chooses. It is suggested that the employer should think carefully before taking such a course of action, and possibly take legal advice. The architect should point out that if the certificate is upheld, the employer will be liable not only for the full amount shown, but also for interest from when the amount should have been paid, and possibly costs in connection with the reference. A more prudent course of action would be to discuss the matter with the architect prior to the next certificate, when an adjustment could be made, although the employer should of course respect that the architect must make an impartial decision.

Interim payment on Practical Completion

7.42   An interim certificate is to be issued not later than 14 days after the date of Practical Completion (cl 4.3(a)), certifying payment of 97.5 per cent of the value referred to in clause 4.2.1(a) and 100 per cent of the value of amounts under clause 4.2.2, together with other deductions as discussed above. The effect of this certificate is to release to the contractor half of the withheld amounts (retention). The employer retains the right to deduct half the percentage from the outstanding amounts due under this certificate.

7.43   There would normally be no further payment certificates between the interim payment at Practical Completion, and the final certificate. However, in some circumstances, for example when a claim for loss and/or expense has not been resolved prior to Practical Completion, the architect should issue a certificate as soon as the claim has been ascertained. This would be exceptional and would apply only to claims or disputed matters, there should be no work instructed or carried out which requires certification following Practical Completion.

Final certificate

7.44   To summarise, by final certificate stage the following certificates should have been issued:

- interim certificates at monthly intervals (cl 4.2(a));
- certificate of Practical Completion (cl 2.9);
- interim certificate following Practical Completion, including release of half of the retention (cl 4.3);
- certificate stating that defects notified have been made good (cl 2.10).

7.45   The final certificate must be issued within the specific time periods set out in the contract, ie within 28 days of sending to the contractor the computations of the adjusted contract sum, or of the certificate stating that defects notified have been made good, whichever is the later.

7.46   The contractor is required to send all documents reasonably required for the adjustment of the contract sum to the architect or the quantity surveyor not later than six months after Practical Completion of the Works (cl 4.5). No later than three months after receiving this information, the architect or quantity surveyor must prepare a statement of all the final valuations of variations and prepare a computation of the contract sum, which will take into account loss and expense, and all other adjustments that have to be made. The statement and computations must be sent to the contractor 'forthwith'. It is worth noting that it has been held that the final certificate can be issued at the same time as the statement, although it would be good practice to allow the contractor time to consider the document *(Penwith District Council v VP Developments)*. The final certificate must state the contract sum as adjusted under clause 4.5. The final certificate can be for a negative amount – in other words it can certify that payment is due from the contractor to the employer.

Penwith District Council v VP Developments Ltd, 21 May 1999, unreported
Penwith employed VP for maintenance works to 91 houses at Hayle. The contract was on JCT80. Practical completion took place on 21 September 1990, and the certificate of making

good defects was issued on 30 October 1991. VP submitted a draft final account on 14 January 1991. Three interim certificates were issued following practical completion, the last one on 10 July 1992. The final certificate was issued on 8 April 1993, and enclosed a document summarising how the figure on the final certificate had been arrived at. VP gave notice of arbitration some three years later. It argued that it was not barred by the clause 30.9 conclusiveness provisions as the final certificate had not been valid. The arbitrator found for VP, stating that the intention of the contract was that the contractor should have at least three months to consider the ascertainment of final account referred to in clause 30.6.1. Penwith appealed and His Honour Judge Humphrey LLoyd found that the contract terms required that no minimum period should have elapsed, all the time limits referred to were maxima. He also found that no such term could be implied, 'the 1980 JCT form is a long and complex document and was plainly intended to provide for most conceivable circumstances and to block the many attempts to find gaps in its structures, despite repeated assaults'.

Conclusive effect of final certificate

7.47   The final certificate is conclusive evidence that proper adjustment has been made to the contract sum, and the contractor is prevented from seeking to raise any further claims for extensions of time, or for reimbursement of direct loss and/or expense (cl 4.7.1). It is also conclusive evidence that where matters have been expressly stated to be for the approval of the architect that they have been approved, but apart from those matters it is not conclusive that any other materials, workmanship, etc are in accordance with the contract.

7.48   Both parties have the right to challenge the issue of the final certificate by commencing proceedings within 28 days (cl 4.7.2), but after this period has elapsed there would be little purpose in either party seeking to raise a claim. The bar on raising matters after the 28-day period cannot be extended by the court, as the bar is an evidential bar and not a bar to bringing arbitration proceedings. In other words arbitration can be commenced, but no evidence can be brought forward.

7.49   The conclusive effect of the final certificate has been the subject of much heated debate over recent years, following the decisions in *Colbart Ltd v H Kumar* and *Crown Estate Commissioners v John Mowlem*. The JCT have amended the relevant clauses in its contracts, so that it should be possible for the employer to bring a claim regarding work or materials which were not in accordance with the contract following the 28-days cut-off period, provided that they had not been stated to be 'to approval' of the architect somewhere within the contract documents. Despite this, the author occasionally comes across practices who, although working with the new forms, are still implementing a policy of never issuing final certificates, or issuing them with alterations or declarations on their face, all measures adopted in the difficult period immediately following the case. There is no longer any justification for taking such steps, and indeed they would constitute a breach of contract on the part of the employer, therefore putting the employer at considerable risk. Provided that the architect has taken all steps as required by the contract, and has made a competent inspection of the Works at the right stage, the certificate must be issued as the contract requires.

Colbart Ltd v H Kumar (1992) 59 BLR 89

Colbart Ltd, a contractor, entered into a contract with Mr Kumar on IFC84 (with Amendments 1 and 2), to carry out work to property of his in south-east London. In November 1990 the architect issued a practical completion certificate, followed one week later by a penultimate certificate, and then three weeks after that by a certificate of making good defects and a final certificate. Three months after the issue of the penultimate certificate, and six weeks after the final certificate, the contractor commenced proceedings for the amounts certified. As part of the defence to this claim, Mr Kumar asserted that some of the work certified was defective. Clause 4.7 of this edition of IFC84 stated that the final certificate was 'conclusive evidence that where and to the extent that the quality of materials or the standard of workmanship are to be to the reasonable satisfaction of the Architect/the Contract Administrator the same are to such satisfaction'. The judge stated that the conclusive effect extended to items where the standard was inherently a matter for the opinion of the architect. Whether or not the quality of any materials or workmanship was inherently a matter for the opinion of the architect was a question of fact and degree in each case. In this project the defective work complained of was such an instance, and there was therefore no arguable defence to the contractor's claim.

Crown Estate Commissioners v John Mowlem & Co Ltd (1994) 70 BLR 1 (CA)

Crown Estates employed Mowlem to construct a commercial development on the site of the former Kensington Palace Barracks. A final certificate was issued on 2 December 1992, and on 6 April 1993, Crown Estates gave notice of arbitration. They then issued a summons under s 27 of the Arbitration Act 1979 for an order extending the time within which to commence arbitration, in order to validate their notice. In addition to the summons the judge at first instance was also asked to consider the question as to what, if anything, the final certificate was conclusive evidence of, as this would affect what could be raised in the arbitration. The judge issued the order extending time and held that the final certificate was only conclusive as to matters that were expressly stated to be for the satisfaction of the architect. Mowlem appealed and the appeal was allowed. The Court of Appeal stated that clause 30.9.1.1 and 30.9.3 did not limit the time within which arbitration proceedings could be brought, therefore the court had no powers under the Arbitration Act 1979 that could defeat the effect of the certificate. It also held that as all standards and quality of work and materials were inherently matters for the opinion of the architect, the final certificate was conclusive evidence of all such matter requires.

8.1   One of the most important functions of a building contract is to establish a clear allocation of liability for the risks inherent in any construction operation, ie the risks of accident, injury and damage to property. Should any unfortunate incidents occur, it is vital that there should be no room for dispute about who is liable for the losses, and that all concerned are clear about what procedural steps must be taken. Ambiguity will only lead to confusion and delays, which will benefit neither party.

8.2   Normally a building contract will set out the specific events for which the contractor is liable, and require the contractor to indemnify the employer in respect of the resultant losses, for example for injury to persons, or damage to neighbouring property. In IFC98 these liabilities are allocated under clause 6.1 and 6.2. Clause 6.1 makes the contractor liable for, and requires indemnification of the employer against, claims for injury to or death of persons, or damage to neighbouring property which has been caused by the contractor's negligence. The indemnity protects the employer in that if an injured party brings an action against the employer, rather than against the contractor, the latter has agreed to carry the consequences of the claim. In practice the employer can either join the contractor as co-defendant or bring separate proceedings against the contractor.

8.3   In practice the indemnities given to the employer by the contractor are quite worthless if the contractor has insufficient resources to meet the claims. IFC98 therefore requires the contractor, under clause 6.2, to carry insurance cover to back up the indemnities required under clause 6.1.

8.4   In addition to the requirement for insurance against claims arising in respect of persons and property, the contract contains alternative provisions for insurance of the Works under clause 6.3. There are also optional provisions requiring the contractor to take out insurance for non-negligent damage to property other than the Works (cl 6.2.4), and for insurance against employer's loss of liquidated damages (cl 6.3D).

Injury to persons and damage to property caused by the negligence of the contractor

8.5   Clause 6.2 requires the contractor to carry insurance to cover injury to persons and damage to property other than the Works, which arise from the carrying out of the Works. The minimum cover required as a contractual obligation is entered in the Appendix. The contractor must be able to provide evidence that this insurance has been taken out. If the contractor defaults, the employer may take out the insurance and deduct the cost from the contract sum.

8.6   The contractor's liability in respect of personal injury or death of employees is met by an employer's liability policy. This has been compulsory since the Employer's Liability (Compulsory Insurance) Act 1969. The contractor's liability in respect of third parties (death or personal injury and loss or damage to property including consequential loss) is met by its public liability policy. Insurers advocate insuring for a minimum of £2,000,000 for any one occurrence, although a minimum cover of £250,000 was all that was required under the Finance (No 2) Act 1975. Liability at common law for claims by third parties is unlimited, and any amount specified in the contract is merely the employer's requirement in the interests of safeguarding against inadequacies, and in no way limits the contractor's liability under clause 6.1.

8.7   Clause 6.2.1 requires that the insurance in respect of personal injury or death of any person in a contract of service with the contractor should comply with 'all relevant legislation'. This would include the insurance required by the Employer's Liability (Compulsory Insurance) Act 1969, and would also cover, eg, insurance requirements under the Road Traffic Act 1988.

8.8   The contractor is required to insure the indemnities required under clause 6.1.1 and 6.1.2 up to the amount stated in the Appendix. However, it is recognised in footnote [p] to clause 6.2.1 that it may not always be possible to acquire insurance cover which is co-extensive with the indemnity required in clause 6.1.1 and 6.1.2. For example, the insurance market has removed gradual pollution from its public liability policies. This in no way affects the contractor's liability and duty to indemnify.

8.9   The liability and duty to indemnify are subject to exceptions. In respect of liability for personal injury or death, this is qualified in that the contractor is not liable where injury or death is caused by an act of the employer, or a person for whom the employer is responsible (cl 6.1.1).

8.10   In respect of damage to property the contractor is only liable to the extent that the damage is caused by negligence or breach of statutory duty or other default of '...the Contractor, his servants or agents, or of any person employed or engaged upon or in connection with the Works ... other than the Employer' (cl 6.1.2). The contractor is therefore liable only for losses caused by its own negligence. It is made clear in clause 6.1.3 and 6.1.4 that the definition of 'property' excludes the Works, up to Practical Completion, except parts taken over by partial possession. The last sentence of clause 6.1.2 excludes liability for loss or damage to property caused by a 'Specified Peril' where this is required to be insured under clause 6.3C.1. This means that, where clause 22C is applicable, the contractor is not liable for losses insured under that clause and caused by a Specified Peril, even where the damage is caused by the contractor's own negligence *(Scottish Special Housing Association v Wimpey Construction)*. Domestic sub-contractors, however, may be liable for losses caused by their negligence *(BT v James Thompson & Sons)*. It should be noted, also, that the contractor might remain liable for some consequential losses *(Kruger Tissue v Frank Galliers)*.

Scottish Special Housing Association v Wimpey Construction UK Ltd (1986) 34 BLR 1
SSHA entered into a contract with Wimpey on JCT63 (1977 edition) to modernise 128 houses in Edinburgh. During the course of the works some of the houses were damaged by fire and it was assumed for the purposes of the case that the fire was caused by the contractor's negligence. Clause 18(2) (the equivalent is clause 6.1.2 in IFC98), which dealt with the contractor's liability for damage to property, was headed by the phrase 'except for such loss and damage as is at the risk of the Employer under ... 20 (C)' (clause 6.3C in IFC98). The court found that this had the effect of exempting the contractor from liability for any loss or damage that was required to be insured under clause 20(C), however caused.

British Telecommunications plc v James Thompson & Sons (Engineers) Ltd
[1999] BLR 35 (HL)
James Thompson were sub-contractors on a refurbishment project for British Telecom being undertaken on a JCT80 form of contract. A fire broke out in the roof area while the sub-contractors were carrying out their work. The court found that the relevant clauses had the same effect as the equivalent clauses considered in *SSHA v Wimpey*. However, it decided that domestic sub-contractors remained under a duty of care to prevent such losses, and were therefore liable to BT under the tort of negligence. The court considered that the wording of clause 22.3, which required the joint names policies to waive the rights of subrogation against nominated but not domestic sub-contractors, should be taken into account in considering whether a duty of care existed. The fact that BT were indemnified by the clause 22C insurers, even if the fire was caused by the sub-contractors, was not sufficient to prevent the imposition of the duty.

Kruger Tissue (Industrial) Ltd v Frank Galliers Ltd (1998) 57 ConLR 1
Damage was caused to existing building and works by fire, assumed for the purposes of
the case to be the result of the negligence of the contractor or sub-contractor. The
construction work being carried out was on a JCT80 form. The employers brought a
claim for loss of profits, increased cost of working and consultants' fees, all of which are
consequential losses. Judge John Hicks decided that the employer's duty to insure for
'the full cost of reinstatement, repair or replacement of the existing structure and the
works under clause 22C (and therefore contractor's exemption from liability under
clause 20.2), did not include such consequential losses'. A claim could therefore be
brought against the contractor for these.

**Damage to property not caused by
the negligence of the contractor**

8.11   The liability for damage to adjoining buildings where there has been no
negligence on the part of the contractor is not covered under clause 6.1.2.
Subsidence or vibration resulting from the carrying out of the Works might
cause such damage, even though the contractor has taken reasonable care.
This is a risk which may be quite high with certain projects on tight urban sites,
or in close proximity to old buildings. In such cases it might be advisable to
take out a special policy for the benefit of the employer.

8.12   In IFC98 there is an optional provision for this type of insurance under clause
6.2.4. If it is anticipated that the main contractor may be required to take out
this insurance, the correct deletion must be made in the Appendix, and the
amount of cover entered. The architect must then instruct the contractor to
take out the policy, after confirming with the employer that the policy is
required. The cost is added to the contract sum. The policy must be in joint
names and placed with insurers approved by the employer. The policy and
receipt are to be deposited with the employer.

8.13   This insurance is usually expensive, and subject to a great many exceptions. If
it is required, then the policy needs to be effective at the start of the site
operations when demolition, excavation, etc are carried out. The text of clause
6.2.4 was revised in 1996 to take account of the wording of model exclusions
compiled by the Association of British Insurers. The policy should be checked
by the employer's insurance advisors to ensure that any exclusions correlate
with clause 6.2.4 and that the policy provides the cover that the clause requires.

Insurance of the Works

8.14   There are three alternative clauses for insuring the Works, and the clause which is applicable should be entered in the Appendix (note there is no need to delete clauses in the form itself). In all cases the policies are to be in joint names, and cover must be maintained up until Practical Completion of the Works, or determination if this should occur earlier. All three are for joint names policies. The 'Joint Names Policy' definition was reworded in 1996 to make clear the intention that, under the policy, the insurer does not have a right of subrogation to recover any of the monies from either of the named parties. The policies must also either cover named sub-contractors or include a waiver of any rights of subrogation against them (cl 6.3.3). With the exception of joint names policies under clause 6.3C.1, the requirements for recognition or waiver apply also to domestic sub-contractors.

8.15   Clause 6.3A and 6.3B cover insuring new building work and require 'All Risks' cover under joint names policies. A definition of 'All Risks' is given in clause 6.3.2 and refers to 'any physical loss or damage to work executed and Site Materials and against the reasonable cost of the removal and disposal of debris …'. There is also a list of exclusions, which includes the cost necessary to repair, replace or rectify property which is defective, loss or damage due to defective design, and loss or damage arising from war and hostilities. Even in a so-called 'All Risks' insurance policy there may be further exclusions, and the employer's insurance advisors should carefully check the wording of each policy. Useful guidance on this and other insurance matters was given in JCT Practice Note 22, written with JCT80 in mind but largely applicable to IFC98.

8.16   Clause 6.3A insurance is taken out by the contractor and is to be for the full reinstatement value of the Works, including professional fees to the extent entered in the Appendix. The contractor is responsible for keeping the Works fully covered, and in the event of under-insurance will be liable for any shortfall in recovery from the insurers.

8.17   Clause 6.3B insurance is taken out by the employer, and again is to be for the full reinstatement value of the Works, including professional fees. The employer is responsible for keeping the Works fully covered, and in the event of under-insurance will be liable for any shortfall.

8.18   Clause 6.3C is applicable where work is being carried out to existing buildings. It includes two insurances, both taken out by the employer. The existing structure and contents must be insured against 'Specified Perils' as defined in clause 8.3 (cl 6.3C.1). New works in, or extensions to, existing buildings must be covered by an 'All Risks' insurance policy (cl 6.3C.2).

The Joint Fire Code

8.19   The Joint Fire Code (cl 6.3FC) is designed to reduce the incidence of fire on construction sites. It is an optional provision, but as compliance with the Code may reduce the cost of some insurance policies, its inclusion should be carefully considered. If included, both parties undertake to comply with the Code and ensure that those employed by them also comply. They indemnify each other against any losses resulting from any breaches on their part of the Code (cl 6.3FC.4).

8.20   If a breach of the Fire Code occurs, the insurers may give notice to either the employer or the contractor of remedial measures they require and the dates by which they must be put into effect. (Note that the provisions in the contract dealing with breaches have recently been revised under Amendment 3.) If either party receives such a notice, they must copy it to the other, and the employer must send a copy to the architect (cl 6.3FC.3.1). If the notice sets out measures which conform with the contractor's existing obligations under the contract, then the contractor should proceed to put the measures in place (cl 6.3FC.1). If the contractor does not comply with the notice within seven days, the employer may employ and pay others to effect such compliance (cl 6.3FC.3.2).

8.21   Where the remedial measures in the notice constitute a variation to the contract, the architect must issue instructions as necessary to enable compliance. In an emergency, ie where the insurer's notice requires immediate compliance, the contractor should take such steps as are reasonably necessary to achieve such compliance, and should inform the architect immediately of the position (cl 6.3FC.2).

Terrorism cover

8.22   The clauses relating to 'All Risks' and 'Specified Perils' include damage caused by fire, explosion etc, and make no distinction between accidental occurance of these events or deliberate cause through acts of terrorism. However, in the aftermath of massive claims arising from damage caused by terrorist activity in London and other cities in the early 1990s, underwriters gave notice that they might withdraw cover, and Amendment TC/94 was published as a supplement to deal with the consequences of such a withdrawal. This supplement has now been incorporated into the form itself by Amendment 3 *(see Appendix D)*. Alongside this amendment the JCT has published Practice Note 3 (series 2), 'Insurance – Terrorism Cover'.

Employer's loss of liquidated damages

8.23   If the contractor is caused delay by one of the Specified Perils, an extension of time would normally be awarded under clause 2.3 and the employer will not be able to claim liquidated damages from the contractor for that period. There will therefore be a loss to the employer as a result. Clause 6.3D.1 is available should the employer wish to have insurance against this loss of liquidated damages. This is an optional clause and conditional upon an entry having been made in the Appendix. There are problems with such insurance as liquidated damages are payable without proof, and traditionally insurers only pay on proof of actual loss. As a result, only one or two firms are currently willing to offer cover, and the price tends to be high.

Other insurance

8.24   There are other insurances not covered by the provisions of IFC98, which the employer might wish to consider. The employer is the party in the best position to assess possible loss. Where there are likely to be business or other economic losses, then these can be insured against, albeit at a price. It is also possible to insure against defects occurring in the buildings by means of project-related insurance. This is still relatively expensive and limited to a ten-year 'decennial' loss. Irrespective of blame, it means that money is available for remedying the defects which will occur most often in the first eight years of the life of a building. Project-related insurance needs to include for subrogation waiver, and in no way reduces the need for professional indemnity cover.

Action following damage to the Works

8.25   The procedure is similar under clause 6.3A.1, 6.3B.1 and 6.3C.1. The contractor must as soon as possible notify the architect and the employer in writing of the details of the damage. Although the contract does not require it, clearly the contractor or employer, depending on who has taken out the policy, should also inform the insurers immediately on becoming aware of the damage. After any inspection required has been made by the insurers the contractor is then obliged to make good the damage and continue with the Works. Under all three clauses, the contractor authorises that all monies due under the insurance policy are paid direct to the employer.

8.26   All three clauses state that 'the occurrence of such loss or damage shall be disregarded in computing any amounts payable to the contractor'. Interim certificates that have already been issued and the amounts paid or due under them are of course not affected by the occurrence of the damage. In addition, any work that was completed after the most recent interim certificate, but was then subsequently damaged, should also be included in the next interim certificate.

8.27   Under clause 6.3A.1 the contractor must take out insurance for the full reinstatement value of the Works, plus a percentage to cover professional fees if this is required in the Appendix. This money, minus the part of it to cover professional fees, should be included in future certificates as the work is carried out. The sums are not added to the contract sum and the reinstatement work is kept distinct by issuing 'reinstatement certificates' which cover only that work, at the same intervals as the normal interim certificates. If the amount paid by the insurers is less than it costs the contractor to rebuild the Works, the contractor is not entitled to any additional payment (cl 6.3A4.5). The risk of any under-insurance therefore lies with the contractor.

8.28   Under clause 6.3B and 6.3C the rebuilding work is treated as if it were a variation under clause 3.6, therefore the contractor is less at risk and the employer will have to bear any shortfall in the monies paid out. Under clause 2.4.3 the contractor is entitled to an extension of time for delay caused by loss or damage due to one or more of the Specified Perils. In addition, as the work is treated as a clause 3.6 variation, the contractor may be entitled to an extension of time and loss and/or expense under clauses 2.4.5 and 4.12.7. In all cases the entitlement appears to extend even to cases where the damage was caused by the contractor's negligence.

8.29   Under clause 6.3C either party is given the right to determine the employment of the contractor 'if it is just and equitable to do so' (cl 6.3C.4.4). This might arise, for example, where an existing structure to which work is being carried out has been completely destroyed, and it would be quite unreasonable to expect the contractor to embark on re-building it. If the contract is determined the provisions of clause 7.11 will apply. It should be noted that this right is in addition to the right under clause 7.13.1 of either party to determine the contract should the Works be suspended for a period of one month due to loss or damage caused by a Specified Peril.

The architect's role in insurance

8.30   The architect has a duty to explain the provisions of the contract to the employer, and should therefore have a working knowledge of insurance matters, although he or she would not be expected to be an expert. The choice of the appropriate option for insuring 'the Works' is particularly important, and advice must be given to the employer concerning the consequences. In addition, the architect, as contract administrator, will monitor the actions of the parties with respect to the insurance clauses, even though, for the most part, the insurance provisions are dealt with directly between the employer and contractor. The architect should be alert to the need for swift action should loss or damage occur, and may be required to attend an inspection by the insurers, or to supply information.

8.31   The employer should take advice from its own insurance experts concerning the suitability and wording of any policies. The architect is primarily a channel of communication, and although a check should be carried out on wording to see that no undesirable exceptions or restrictions exist that might affect the carrying out of the Works, the main responsibility should rest with the employer and the employer's broker or insurance advisors.

8.32   Where the insurance requirements of the contract cannot be matched by effective cover, then the employer should seek expert advice. For example, the building might be special and uninsurable, or the employer might not wish to have insurance, etc. Decisions in such situations will also have implications for contractors and sub-contractors, and expert advice must be sought.

9.1   Given the complexity and unpredictability of construction operations, it would be unlikely that a project could proceed to completion without breaches of the contractual terms by one party or another. This is recognised by most construction forms, which usually include provisions to deal with forseeable situations. These provisions avoid arguments developing or the need to bring legal proceedings as the parties have agreed in advance a machinery for dealing with the breach. A clear example of this is the provisions for liquidated damages – the contractor is technically in breach if the project is not completed by the contractual date, but all the consequences and procedures for dealing with this are set out in the contract itself. However, some breaches may have such significant consequences that the other party may prefer not to continue with the contract, and for these more serious breaches the contract contains provisions allowing the determination of the employment of the contractor.

Repudiation or determination

9.2   In any contract, where the behaviour of one party makes it difficult or impossible for the other to carry out the obligations of the contract, the injured party might allege prevention of performance and sue either for damages or a quantum meruit. This could occur in construction, eg, where the employer refuses to allow access to part of the site.

9.3   Where it is impossible to expect further performance from a party, then the injured party may claim that the contract has been repudiated. Repudiation is when one party makes it clear that they no longer intend to be bound by the provisions of the contract. This might be expressly stated, or implied by the party's behaviour.

9.4   Most JCT contracts include determination clauses, which provide for the effective termination of the employment of the contractor in circumstances which may amount to, or which may fall short of, repudiation. It should be noted that the determination is of the contractor's employment under the contract, and is not determination of the contract itself. This means that the parties remain bound by its provisions, and can bring actions for losses suffered through breach of its terms.

9.5   If there is repudiation, invoking a determination clause is unnecessary, because the injured party can accept the repudiation and bring the contract to an end. However, the determination provisions are useful in setting out the exact circumstances, procedures and consequences of the termination of employment. These procedures must be followed with great caution because if they are not administered strictly in accordance with the terms of the contract, this in itself could amount to a repudiation of the contract. This, in turn, might give the other party the right to treat the contract as at an end and claim damages.

9.6   Determination can be initiated by the employer (cl 7.2) in the event of specified defaults by the contractor such as suspending the works or failing to comply with the CDM Regulations, or in the event of the insolvency of the contractor. Determination can be initiated by the contractor (cl 7.9) in the event of specified defaults by the employer such as failure to pay the amount due on a certificate, or where specified events result in the suspension of work beyond a period to be entered in the Appendix. Determination might also follow the insolvency of the employer. In the event of neutral causes, which bring about the suspension of the uncompleted Works for the period listed in the Appendix, determination can be exercised by either party (cl 7.13).

Determination by the employer

9.7   The contract provides for determination of the employment of the contractor under stated circumstances. IFC98 expressly states that the right to determine the contractor's employment is 'without prejudice to any other rights or remedies' (cl 7.8). This determination can be initiated by the employer in the event of specified defaults by the contractor occurring prior to Practical Completion (cl 7.2.1), the insolvency of the contractor (cl 7.3.1), or corruption (cl 7.4). In the case of bankruptcy of the contractor (cl 7.3.3) the determination is automatic.

9.8   The procedures as set out in the contract must be followed exactly, especially those concerning the issue of notices (see figure 10). If default occurs the architect should issue a warning notice (cl 7.2.2). If the default continues for 14 days from receipt of the notice, then the employer may determine the employment of the contractor by the issue of a further notice within ten days from the expiry of the 14 days. If the contractor ends the default or if the employer gives no further notice, and the contractor then repeats the default, the employer may determine 'within a reasonable time after such repetition' (cl 7.2.3). The employer must still give a notice of determination, but no further warning is required from the architect. There appears to be no time limit on the repetition of the default. It should be noted that to be valid, all notices must be in writing and given by actual delivery, or by special or recorded delivery (cl 7.1). This rules out e-mail or fax transmissions. As time limits are of vital importance it is also wise to have receipt of delivery confirmed.

9.9   The grounds for determination by the employer must be clearly established and expressed. The contract clearly states that determination must not be exercised unreasonably or vexatiously (cl 7.2.4). Under clause 7.2.1(a)

Figure 10 Determination

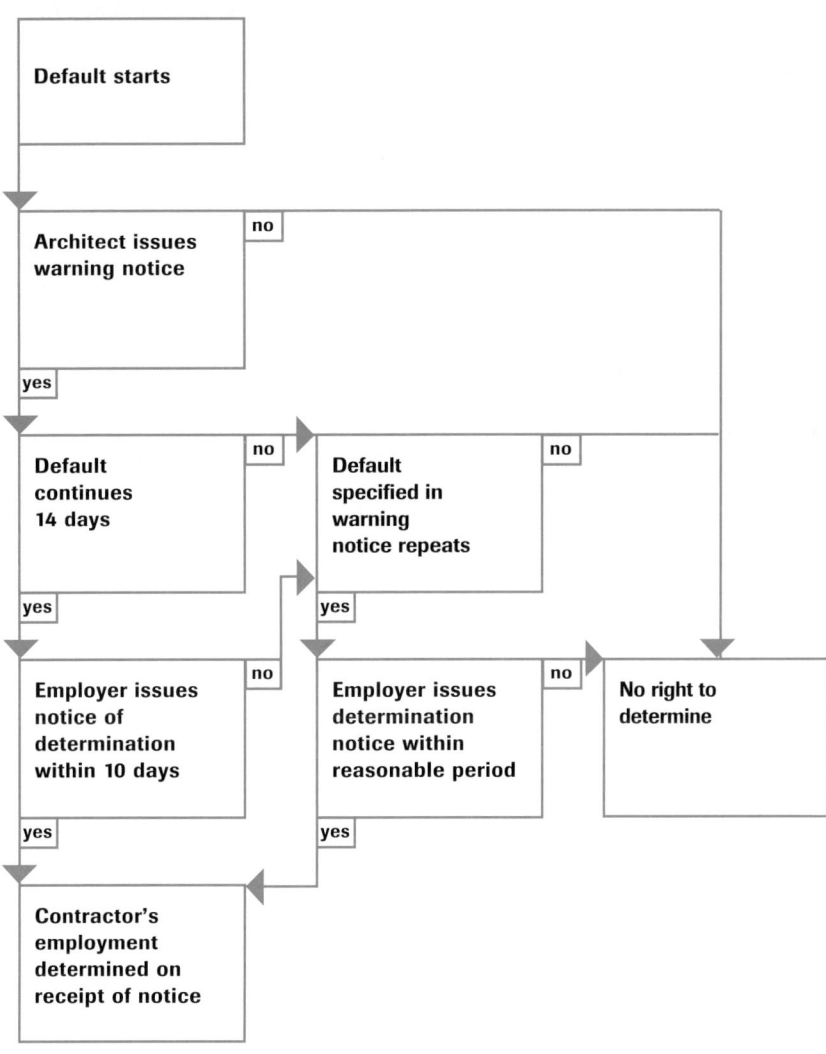

suspension of the work must be whole and substantial, and 'without reasonable cause'. However, the contractor might find 'reasonable cause' in any of the matters referred to in clause 4.12. An exercise of the right to suspend work under clause 4.4A would not be cause for determination, provided that it had been exercised in accordance with the terms of the contract.

Specified defaults

9.10 The specified defaults which may give rise to determination are that the contractor: wholly or substantially suspends the carrying out of the works without reasonable cause; 'fails to proceed regularly and diligently'; refuses or neglects to comply with a written instruction requiring the contractor to remove defective work, fails to comply with clause 1.3 (assignment), clause 3.2 (sub-contracting) and clause 3.3 (named persons); and fails to comply with the contractual provisions of the CDM Regulations. Generally speaking, the default would have to be serious to justify determination, although any failure to comply with the CDM provisions which would put the employer at risk of action by the authorities would be sufficient.

9.11 The default that the contractor 'fails to proceed regularly and diligently' (cl 7.2.1(b)) is notoriously difficult to establish, and although meticulous records will help, architects are often understandably reluctant to issue the first warning notice. It means more than simply falling behind any submitted programme, even to such an extent that it is quite clear the project will finish considerably behind time. However, something less than a complete cessation of work on site would be sufficient grounds.

9.12 In the case of *London Borough of Hounslow v Twickenham Garden Developments*, eg, the architect's notice was heavily attacked by the defendants. In a more recent case, however, the architect was found negligent because he failed to issue a notice *(West Faulkner Associates v London Borough of Newham)*. It should be remembered that without the first 'warning notice' issued by the architect the employer cannot issue the determination notice.

London Borough of Hounslow v Twickenham Garden Developments (1970) 7 BLR 81
The London Borough of Hounslow entered into a contract with Twickenham Garden Developments to carry out sub-structure works at Heston and Isleworth in Middlesex. The contract was on JCT63. Work on the contract stopped for approximately eight months due to a strike. After work resumed the architects issued a notice of default stating that the contractor had failed to proceed regularly and diligently and that unless there was an appreciable improvement the contract would be determined. The employers then proceeded to determine

the contractor's employment. The contractor disputed the validity of the notices and the determination, and refused to stop work and leave the site. The council applied to the court for an injunction to remove the contractor. The judge emphasised that an injunction was a serious remedy and that before he could grant one there had to be clear and indisputable evidence of the merits of their case. The evidence put before him, which showed a significant drop in the amounts of monthly certificates and numbers of workman on site, failed to provide this.

West Faulkner Associates v London Borough of Newham (1992) 61 BLR 81

West Faulkner were architects engaged by the council for the refurbishment of a housing estate consisting of several blocks of flats. The residents of the estate were evacuated from their flats in stages to make way for the contractors, Moss, whom it had been agreed would carry out the work according to a programme of phased possession and completion, with each block to take nine weeks. Moss fell behind the programme almost immediately. However, they had a large workforce on the site and continually promised to revise their programme and working methods to address the problems of lateness, poor quality work and unsafe working practices that were drawn to their attention on numerous occasions by the architects. In reality Moss remained completely disorganised, and there was no apparent improvement. The architects took the advice of quantity surveyors that the grounds of failing to proceed regularly and diligently would be difficult to prove, and decided not to issue a notice. As a consequence the council were unable to issue a notice of determination, had to negotiate a settlement with the contractors and dismissed the architects, who then brought a claim for their fees. The judge decided that the architects were in breach of contract in failing to give proper consideration to the use of the determination provisions. In his judgment he stated that '"regularly and diligently" should be construed together and in essence they mean simply that the contractors must go about their work in such a way as to achieve their contractual obligations. This requires them to plan their work, to lead and manage their workforce, to provide sufficient and proper materials and to employ competent tradesmen, so that the works are carried out to an acceptable standard and that all time, sequence and other provisions are fulfilled' (Judge Newey at page 139).

Insolvency of the contractor

9.13   Insolvency is the inability to pay debts as they become due for payment. Insolvent individuals may be declared bankrupt. Insolvent companies may be dealt with in a number of ways depending upon the circumstances, eg by voluntary liquidation (in which the company resolves to wind itself up); compulsory liquidation (under which the company is wound up by a court order); administrative receivership (a procedure to assist the rescue of a company under appointed receivers); an administration order (a court order given in response to a petition, again with the aim of rescue rather than liquidation, and managed by an appointed receiver); or voluntary arrangement (in which the company agrees terms with creditors over payment of debts). Procedures for dealing with insolvency are mainly subject to the Insolvency Act 1986 and the Insolvency Rules 1996 (SI 1996/1925). Under these the person authorised to oversee statutory insolvency procedures is termed an insolvency practitioner.

9.14   Under IFC98, the contractor must notify the employer in writing in the event of liquidation or insolvency (cl 7.3.2). The contract provides for automatic determination in the event of the contractor's liquidation, ie where the contractor becomes bankrupt, has a winding up order made, passes a resolution for a voluntary winding up, or has a provisional liquidator or trustee in bankruptcy appointed. There is an option for reinstatement, but this would, of course, be a matter for the employer and the insolvency practitioner.

9.15   Determination is not automatic, however, where the contractor makes an arrangement with his creditors, or makes a proposal for a voluntary composition of debts or scheme of management under the Companies Act 1985 or the Insolvency Act 1996, or where an administrator or administrative receiver is appointed. In these cases the employer is given an option to determine (cl 7.3.4), or to consider a more constructive approach. This is to allow the appointed insolvency practitioner time to come up with a rescue package if possible. It is usually in the employer's interest to have the Works completed with as little additional delay and cost as possible, and a breathing space might allow all possibilities to be explored. During this period the employer is under no obligation to make further payment, and the contractor is relieved of the obligation to 'carry out and complete the Works' (cl 7.5.1). The employer may make reasonable arrangements to protect the site, and the contractor must not hinder this. The employer can then either make an agreement under clause 7.5.2.1 to arrange for the work to continue, or determine the employment of the contractor.

9.16   A '7.5.2.1' agreement basically provides three options. The first is that arrangements may be made for the contractor to continue and complete the Works. Unless the insolvency practitioner has been able to arrange resource backing, this may not be a realistic option. If Practical Completion is near, however, and there is money due to the contractor, it can be advantageous to allow completion under the insolvency practitioner's control.

9.17   As an alternative, another contractor may be novated to complete the Works. On a 'true novation', the substitute contractor would take over all the original obligations and benefits (including completion to time and within contract sum). More likely is the third option which is 'conditional novation' whereby the contract completion date, etc would be subject to re-negotiation, and the substitute contractor would probably want to disclaim liability for that part of the work undertaken by the original contractor.

9.18 Deciding on which of the options would best serve the interests of all the parties is a matter between the employer, no doubt advised by the architect, and the insolvency practitioner. There might be merit in adopting one particular course of action, or there might be advantages in taking a more pragmatic approach. For example it may prove expeditious to continue initially with the original contractor under an interim arrangement (cl 7.5.3) until such time as novation can be arranged, or a completion contract negotiated.

Consequences of determination by the employer

9.19 If the contract is automatically determined, or if the employer exercises its right to determine, then the employer may employ another contractor to complete the Works (cl 7.6(b)), or elect not to complete the Works (cl 7.7.1). The original contractor is required to give up possession of the site (cl 7.6(a)) and must remove temporary plant, etc when required (including arranging for plant, etc not owned by the contractor to be removed by the owner). Otherwise the employer may do so (cl 7.6(e)).

9.20 The employer has the right to:
- use any temporary buildings, plant etc on site which are not owned by the original contractor, subject to the consent of the owner (cl 7.6(b));
- require the original contractor to assign the benefit of any sub-contracts to employer (except if insolvent or in liquidation, and to the extent that the benefit is assignable) (cl 7.6(c));
- pay any sub-contractor (except where the main contractor is in liquidation) (cl 7.6(d)) *(see also B Mullan v John Ross)*.

B Mullan & Sons Contractors v John Ross (1996) 86 BLR 1
Mullan was a sub-contractor to McLaughlin and Harvey for the construction of a transit shed. There was no written sub-contract and the main contract was on JCT WCD 81. After the sub-contractor had completed the works and had received an interim payment, an amount of around £114,000 was still outstanding. In October 1993 the main contractor was placed in administrative receivership. Mullan applied to the employer for direct payment under clause 27 of the main contract. The employer indicated that it was willing to pay, but first wished to establish the amounts owing to all sub-contractors. In January

1994 a resolution for the voluntary winding up of the main contractor was passed at an extraordinary general meeting of that company and the defendant was appointed one of the joint liquidators. Mullan applied to the court for directions as to whether the employer could make Mullan a payment. It was held that the employer could not make payments direct, but that all payments must be to the receivers, as otherwise this would be contrary to the pari passu rule whereby assets should be shared in proportion between creditors. The decision appears to narrow the ambit of clause 7.6(d), and the circumstances in which the employer may make direct payment to sub-contractor following determination.

9.21   If the employer decides to employ others under clause 7.6(b), this must be handled quite carefully, as completion of a building started by another contractor is always difficult A completion contract might result from negotiation or competitive tender, but the latter may be advisable if there is much to complete as later the employer may have to justify that the costs incurred were reasonable. A record should be made of the exact state of completeness at the time of determination, incuding any defective work.

9.22   The contractor is entitled to payments of amounts properly due up to 28 days before the date of determination. In situations where the employer had the right to determine, this would be 28 days before the date when it could first give notice (cl 7.6(f)). A notional final account must be set out stating what is owed or owing, either in a statement prepared by the employer or in an architect's certificate (cl 7.6(g)). The employer is entitled to all direct losses caused by the determination, and it is more than likely that the account will result in a claim agianst the contractor.

9.23   One of the consequences of determination is that it often takes time for the contractor to effect an orderly withdrawal from site, and for the employer to establish the amounts outstanding before final payment. Should the employer decide not to continue with the construction of the Works after determination, the employer is required to notify the contractor in writing within six months of that notice (cl 7.7.1). If after that time the contractor has not received written notice, and has not received the employer's statement of account, the contractor may require the employer to state in writing its intentions (cl 7.7.2).

Determination by the contractor

9.24   The contractor is given a reciprocal right to determine its own employment in the event of specified defaults of the employer (cl 7.9.4), or specified suspension events (cl 7.9.2), or insolvency of the employer (cl 7.10). The specified events must have resulted in the suspension of the whole of the uncompleted Works for the continuous period stated in the Appendix. In the case of specified defaults or suspension events a notice is required, which must specify the default or event. If the default or event continues for 14 days from receipt of the notice, the contractor may determine the employment by a further notice up to ten days from the expiry of the 14 days. Alternatively, if the employer ends the default or the suspension event ceases, and the contractor gives no further notice, should the employer repeat the default the contractor may determine 'within a reasonable time after such repetition' (cl 7.9.4). These notices must be given by actual delivery, special delivery or recorded delivery, and not by fax or e-mail.

9.25   The grounds differ from those that give the employer the right to determine. They include failure to pay an amount properly due on a certificate, obstruction of the issue of a certificate and failure to comply with the contractual provisions relating to the CDM Regulations. There are also matters which relate directly to the duties of the architect, where for example the carrying out of the whole or substantially the whole of the Works is suspended for a period of one month, due to the contractor not having received in time necessary instructions and drawings (cl 7.9.2), or as a consequence of an instruction relating to inconsistencies, variations, or postponement, but only if the instruction was not required through some negligence or default of the contractor. The contractor may also determine if the Works are suspended through delay by persons engaged directly by the employer, or failure by the employer to give ingress to or egress from the site.

9.26   Determination by the contractor is optional in the case of the employer's bankruptcy or insolvency (cl 7.10.3). The contractor must issue a notice and determination would take effect from the receipt of the notice.

Consequences of determination by the contractor

9.27   The contractor must then remove from the site all temporary buildings, tools, etc (cl 7.11.1). Within 28 days of determination the contractor is to be paid the retention deducted prior to the determination (cl 7.11.2). The contractor then prepares an account setting out the total value of the work at the date of determination, plus other costs relating to the determination as set out in clause 7.11.3. These may include such items as the cost of removal and any direct loss and/or damage consequent upon determination (cl 7.11.3(c) and (d)). The contractor is in effect indemnified against any damages that may be caused as a result of the determination. This would not necessarily be the case if the contractor did not comply with the contractual provisions; in that case it might constitute repudiation.

Determination by either the employer or the contractor

9.28　　Either party is given the right to determine if the carrying out of the Works is wholly or substantially suspended for three months due to force majeur, loss or damage to the Works caused the Specified Perils, or civil commotion, or for one month due to one or more of the events listed in clause 7.13.1(d)–(f). These include architect's instructions issued as a result of negligence or default of a local authority or statutory undertaker executing work solely in pursuance of its statutory obligations. The right of the contractor to determine in the event of a Specified Peril is limited by the proviso that the event must not have been caused by the contractor's negligence (cl 7.13.2).

9.29　　Notice may be given by either party and the employment of the contractor will be determined seven days after receipt of the notice, unless the suspension is terminated within that period (cl 7.13.1). Detailed provisions are set out regarding the consequences of the determination. The contractor must remove all temporary buildings, tools etc from the site. The contractor is then paid one half of any retention deducted prior to the determination. Within two months of determination the contractor must provide the employer with all the necessary documents for the purposes of preparing an account. The items that are to be considered for inclusion in the account are set out in clause 7.18.1 to 7.18.4. If determination is due to an event caused by a Specified Peril, which was in turn caused by the negligence or default of the employer, then the contractor may be due an item for direct loss and/or damage suffered as a result of the determination (cl 7.19).

Determination of the employment of a named person

9.30　　The contractor is responsible for taking action with respect to determination of the employment of named persons as sub-contractors. The contract states that the employment of a named person must not be determined other than through operation of clauses 27.2, 27.3 or 27.4 of NAM/SC, and that the contractor must not bring the sub-contract to an end through acceptance of the repudiation of the sub-contract (cl 3.3.3). The reason for this requirement is that once the sub-contract has been determined, the contractor is required to take steps to recover from the sub-contractor any additional amounts payable to the contractor by the employer as a result of the determination (cl 3.5.6(b)). If the contractor has not followed the provisions of NSC exactly, the chances of recovery of these losses would be greatly reduced.

9.31　　The contractor must advise the architect of any events which might give rise to determination of the sub-contract. In some circumstances this may give the architect the opportunity to make some investigations and assess in advance the possible alternative courses of action should determination occur. The contractor must notify the architect as soon as the contract has been determined. The architect must then issue instructions which may either name another person to execute the work, require the contractor to complete the work, or omit the outstanding work (cl 3.3.3(a), (b) and (c) respectively). The consequences of the determination then depend on whether the sub-contractor was originally named in the contract documents, or named in an instruction relating to a provisional sum.

9.32   If the sub-contractor was originally named in the tender documents, then an instruction naming a replacement is treated as an event which may be grounds for an extension of time, but not as a 'matter' giving rise to direct loss and/or expense (cl 3.3.4). The contract sum is to be adjusted by the difference between the price of the first named sub-contractor for the outstanding work, and the price of the replacement, although any amounts in the price of the replacement sub-contractor which cover the correction of defective work are not to be added to the contract sum. The effect of this is that the contractor remains responsible for any defective work carried out by the original named person. An instruction omitting the work, or requiring the contractor to carry out the work, is treated as a variation under clause 3.7, and an event which may give rise to an extension of time and to an award of direct loss and/or expense.

9.33   If the determination relates to a sub-contractor named under an instruction relating to a provisional sum, then the architect's instruction under clause 3.3.3(a), (b) or (c) is treated as a further instruction under the provisional sum, and therefore one which may give rise to an adjustment of the contract sum, an extension of time and to an award of direct loss and/or expense (cl 3.3.5). The architect should be careful not to delay unreasonably in issuing instructions following a determination, as this could also be grounds for a claim for extension of time under clause 2.4.7, direct loss and/or expense under clause 4.12.1, and under exceptional circumstances could even lead to determination under clause 7.9.2(a).

9.34   If the employment of a named sub-contractor is determined other than in accordance with the contractual provisions, then the architect is still required to issue instructions as described above, but the contract states that these will not result in any right to an extension of time or to direct loss and/or expense, and that no adjustment will be made to the contract sum except if application of the contractual provisions would result in a reduction (cl 3.3.6). The effect of this is to place the entire risk of the consequences of such a determination on the shoulders of the contractor. By failing to comply with the terms of the contract, the contractor may have placed the employer in a position where no losses can be recovered from the sub-contractor, and therefore these losses are borne by the contractor.

10.1   IFC98 refers to four methods of dispute resolution: mediation; adjudication; arbitration; and legal proceedings. One of these, adjudication, is a statutory right, and if one party wishes to use this method the other has no alternative but to concur. Mediation is a voluntary process and therefore its use depends on the agreement of both parties and their cooperation throughout the process. If neither of these processes is used, or if either party is dissatisfied with the decision of an adjudicator, then the dispute will have to be determined by arbitration or litigation. IFC98 requires the parties to decide in advance which of these processes will be used.

10.2   There are therefore stages, either before or during the contract, where the parties have the opportunity to agree a preferred course of action. It is important for the architect to understand and to be able to advise on these methods. In addition the architect ought to be familiar with the processes as they may find that they become involved in giving evidence. However, the architect should be careful to avoid giving advice about the merits of a case, or how the employer should bring or defend a claim, as this is normally beyond the architect's expertise, although of course some general advice and background information may be of some help.

10.3   The architect should also tread carefully if the employer decides to attempt to resolve differences through negotiation. This might be the best solution to the problem, but the architect has no authority to negotiate amendments to the terms of the contract or make ad hoc agreements on behalf of the employer. Even if the employer gives the architect an extended authority to negotiate a settlement, where the dispute involves complex legal points, a lawyer would best handle the negotiations.

Alternative dispute resolution

10.4   If negotiations fail to achieve an agreement, the parties may submit the dispute to 'alternative dispute resolution' (ADR), a name used to cover methods such as conciliation, mediation, and the mini-trial. IFC98 draws attention to the possibility of using mediation in footnote [uu] to Part 4: Settlement of disputes, which refers to JCT Practice Note 28 *Mediation on a Building Contract or Sub-Contract Dispute (see Appendix E)*.

10.5   This useful Practice Note sets out the procedure that may be used in mediation. A mediator is appointed jointly by the parties, and will normally meet with the parties together and separately, in an attempt to resolve the differences. The outcome is in the form of a recommendation which, if acceptable, can be signed as a legally binding agreement. This would then be enforceable in the same way as any other contract. However, if the recommendation is not acceptable to one of the parties and is not signed as a binding agreement, it cannot be imposed by law, and so the time spent on the mediation may appear to have been wasted.

10.6   Nevertheless, there can be many advantages to mediation. Unlike adjudication, arbitration or litigation, it is a non-adversarial process which tends to forge good relationships between the parties. Imposed solutions may leave at least one of the parties dissatisfied and may make it very difficult to work together in the future. If the parties are keen to promote a long-term business relationship they should give mediation serious consideration. Even if mediation does not result in a complete solution it has been found in practice that it can help to clear the air on some of the issues involved and establishes common ground. This, in turn, might then pave the way for shorter and possibly less acrimonious arbitration or litigation.

Adjudication

10.7   The Housing Grants, Construction and Regeneration Act 1996 requires that parties to the construction contracts falling within the definition set out in the Act have the right to refer any dispute to adjudication. Article 5 of JCT98 re-states this right, and the related provisions are set out in clause 41A. The JCT provisions comply with the 1996 Act but also introduce requirements over and above those imposed by statute.

10.8   The adjudicator may be either named in the contract, agreed, or nominated. If the adjudicator is named, specific amendments must be made to clause 9A, which are set out in JCT Practice Note 2 (series 2). A named adjudicator enters into the JCT Adjudication Agreement (named) with the parties at the time the main contract is entered into.

10.9   The party wishing to refer the dispute to adjudication must give notice under clause 9A.4.1. The notice should identify briefly the dispute or difference. If no adjudicator is named the parties may either agree an adjudicator or either party may apply to a 'nominator', who is selected in advance from a list of organisations set out in the Appendix. If no nominator is selected, then the contract states that the nominator will be the RIBA. The adjudicator should be appointed on the JCT Adjudication Agreement **(cl 9A.3). (See figure 11)**.

10.10   The JCT provisions do not require any qualifications of the adjudicator but as they refer to the adjudicator as an 'individual' (distinguished from a partnership or company in the definition of 'person' under clause 8.3), it should be an

Figure 11 Dispute resolution

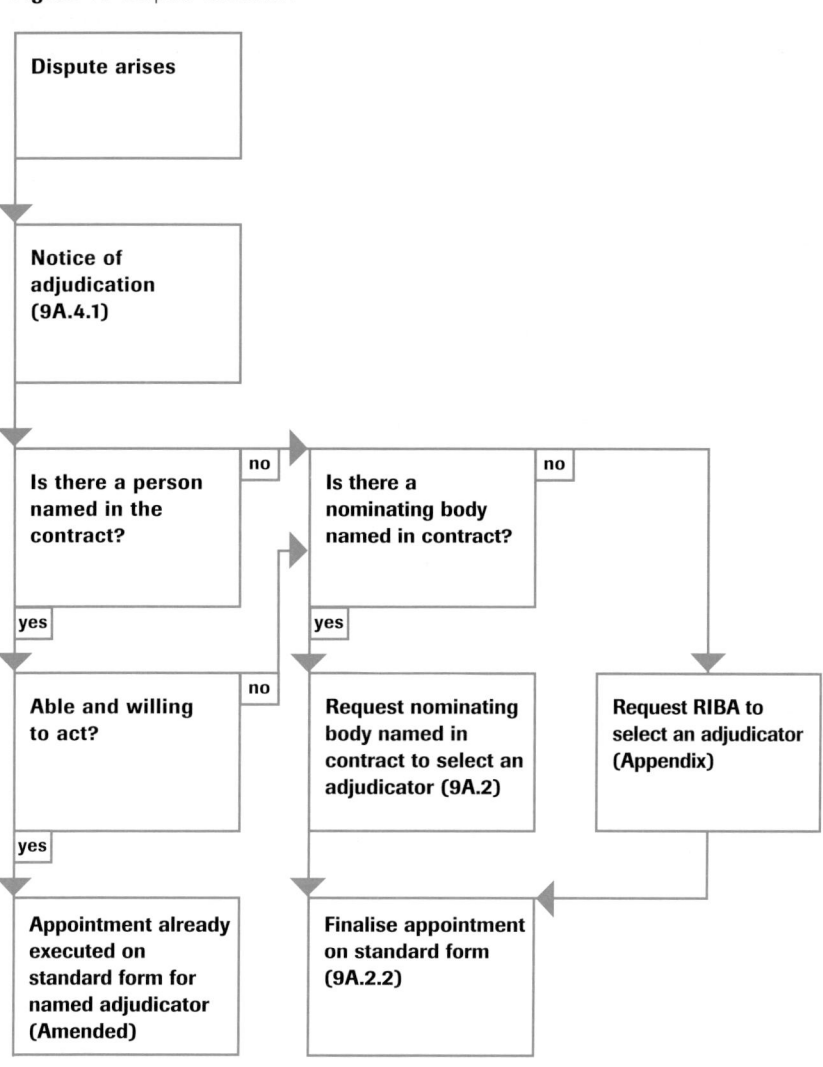

appointment of a natural person acting in a personal capacity. It should be noted that the nominators listed in IFC98 have agreed with the JCT that they will only nominate persons who are prepared to enter into the JCT Adjudication Agreement, therefore an adjudicator put forward through that route should not be insisting on special terms.

10.11   The 1996 Act requires that the reference to the adjudicator is made 'within seven days of the Notice'. The JCT provisions state 'within 7 days of such notice or … immediately upon appointment' (cl 9A.4.1).

10.12   The referral must include particulars of the dispute, a summary of the contentions on which the referring party relies, a statement of relief or remedy sought, and any material it wishes the adjudicator to consider (cl 9A.4.1). The adjudicator must confirm receipt of referral documents (cl 9A.5.1).

10.13   The adjudicator must then set out the procedure to be followed (cl 9A.5.5). A preliminary meeting may be held to discuss this, otherwise the adjudicator may send the procedure and timetable to both parties. The party who did not initiate the adjudication (the responding party) is required to respond within seven days of the date of referral (cl 9.5.2). The adjudicator is likely to hold a short hearing of a few days at which the parties can put forward further arguments and evidence. There may also be a site visit. Often it is possible to do the whole thing by correspondence (often termed 'documents only').

10.14   The adjudicator is given various powers under clause 9A.5.5, including the right to use his or her own expertise in deciding the dispute, the right to revise decisions and certificates of the architect, the right to request the parties to carry out tests, and the right to obtain advice. The adjudicator must give advance notice if he or she intends to take legal or technical advice. In addition the provisions require the adjudicator to include an estimate of the likely cost (cl 9A.5.5.7).

10.15   The Act requires that the decision is reached within 28 days of referral, but it does not state how this date is to be established (cl 108(2)(c)). The JCT provisions define the period as 28 days from the date of receipt of the referral and its accompanying documentation (cl 9A.5.3). The decision must be delivered forthwith to the parties, so the adjudicator may not retain it pending payment of his or her fee. The provisions state that the adjudicator shall not be required to give reasons for the decision. This does not prevent the adjudicator agreeing with the parties that reasons shall be given; it simply means that they cannot insist on them.

10.16 The adjudicator has no power to award costs. The only exception is the adjudicator's own fees and expenses, plus the adjudicator may order one party to pay the costs of any tests or opening up that he or she has required. The adjudicator will be entitled to charge fees and expenses, although expenses are limited to those 'reasonably incurred'. The adjudicator is required to apportion those fees between the parties. If the adjudicator fails to apportion the fees and expenses, the parties will bear these in equal proportions.

10.17 The parties are jointly and severally liable to the adjudicator for his or her fees and expenses. This means that even if the fees have been apportioned between the parties as discussed above, the adjudicator can claim the entire fees from either party and that party will then have to reclaim the difference from the other party.

10.18 The parties may revoke the appointment of the adjudicator by agreement. The adjudicator is entitled to any reasonable amount of fees and expenses as he or she may determine. The adjudicator may also (by implication) determine how these are to be apportioned; the parties are jointly and severally liable for any amount where the appointment has not been determined as discussed above. If the adjudicator fails to reach a decision in the prescribed time, the adjudicator will not be entitled to any fees and expenses, provided that the parties revoke the adjudicator's appointment.

Arbitration

10.19 Arbitration refers to proceedings in which the arbitrator has power derived from a written agreement between the parties to a contract, and which is subject to the provisions of the Arbitration Act 1996. Arbitration awards are enforceable at law. An arbitrator's award can be subject to appeal on limited grounds.

10.20 If arbitration is preferred to litigation as the method for final determination of disputes, then this is confirmed by selecting Article 7A. The arbitration provisions are set out in clause 41B, which refer to the Construction Industry Model Arbitration Rules ('the Rules'). The Arbitration Act 1996 confers wide powers on the arbitrator unless the parties have agreed otherwise, but leaves detailed procedural matters to be agreed between the parties or, if not so agreed, to be decided by the arbitrator. To avoid problems arising, it is advisable to agree as much as possible of the procedural matters in advance, and IFC98 does this by incorporating the Rules, which are very clearly written and self-explanatory. The specific edition referred to is the 1998 edition published by the JCT, which amends some of the Rules and incorporates supplementary and advisory procedures. As the Rules are likely to be subject to some adjustment in the near future, only a broad outline is given below.

10.21   The party wishing to refer the dispute to arbitration must give notice as required by IFC98 clause 41B.1.1 and Rule 2.1, identifying briefly the dispute and requiring the party to agree to the appointment of an arbitrator. If the parties fail to agree within 14 days, either party may apply to the 'appointor', selected in advance from a list of organisations set out in the Appendix. If no appointor is selected, then the contract states that the appointor will be the RIBA. Under Rule 2.5 the arbitrator's appointment takes effect when he or she agrees to act, and is not subject to first reaching agreement with the parties on matters such as fees.

10.22   The arbitrator has the right and the duty to decide all procedural matters, subject to the parties' right to agree any matter (Rule 5.1). Within 14 days of appointment the parties must each send the arbitrator and each other a note indicating the nature of the dispute and amounts in issue, the estimated length for the hearing, if necessary, and the procedures to be followed (Rule 6.2, as amended). The arbitrator must hold a preliminary meeting within 21 days of appointment to discuss these matters (Rule 6.3 as amended). The first decision to make is whether Rule 7 (short hearing), Rule 8 (documents only), or Rule 9 (full procedure) is to apply. The decision will depend on the scale and type of dispute.

10.23   Under all three Rules referred to above, the parties exchange statements of claim and of defence, together with copies of documents and witness statements on which they intend to rely. Under Rule 8 the arbitrator makes his or her award based on the documentary evidence only. Under Rule 9 the arbitrator will hold a hearing at which the parties or their representatives can put forward further arguments and evidence. There may also be a site visit. The JCT Amendments set out time limits for these procedures.

10.24   Under Rule 7 (as amended) a hearing is to be held within 21 days of the date when Rule 7 becomes applicable, and the parties must exchange documents not later than seven days prior to the hearing. The hearing should be not more than one day. The arbitrator publishes the award within one month of the hearing. The parties bear their own costs.

10.25   The arbitrator is given a wide range of powers under Rule 4, including the power to obtain advice (Rule 4.2), the powers set out in section 38 of the Arbitration Act 1996 (Rule 4.3), the power to order the preservation of work, goods and materials even though they are a part of work that is continuing

(Rule 4.4), the power to request the parties to carry out tests (Rule 4.5), and the power to award costs. Under clause 9B.2 of IFC98 the arbitrator is also given wide powers to review and revise any certificate, opinion, decision, requirement or notice and to disregard them if need be, where seeking to determine all matters in dispute.

10.26   Where the arbitrator has the power to award costs, this will normally be done on a judicial basis, ie the loser will pay the winner's costs (Rule 13.1). The arbitrator will be entitled to charge fees and expenses and will apportion those fees between the parties on the same basis. The parties are jointly and severally liable to the arbitrator for fees and expenses incurred.

Arbitration and adjudication

10.27   Under Article 8 any dispute that has been referred to an adjudicator may be referred to arbitration if either party requires this. Clause 4.7.2 states that even where the decision has been given after the final certificate is issued, either party may refer the dispute to arbitration, provided the arbitration is commenced within 28 days of the adjudicator's decision. It is not entirely clear, however, whether this will affect the date at which the final certificate will become conclusive evidence of the matters listed under clause 4.7.1, which may of course have been the matters disputed under the adjudication.

Arbitration or litigation

10.28   As stated above, IFC98 contains alternative provisions for arbitration and litigation in Articles 9A and 9B, and a choice has to be made before tender documents are sent out. Both processes give rise to binding and enforceable decisions. Both tend to be lengthy and expensive, although there are provisions for short forms of arbitration.

10.29   Article 9B confirms that where reference to the arbitration clause has been deleted, then the method for finally resolving disputes is to be by litigation, or 'legal proceedings' as it is referred to in IFC98. The very brief clause 9C would then apply.

10.30   Litigation cases involving claims for amounts greater than £25,000 are normally heard in the High Court, and construction cases are usually heard in the Technology and Construction Court, a specialist department of the High Court which deals with technical or scientific cases. Procedures in court follow the Civil Procedure Rules, with the timetable and other detailed arrangements being determined by the court. A judge will hear the case, and in the High Court a barrister must represent the parties.

10.31   Disputes in building contracts have traditionally been settled by arbitration. Arbitrators are usually senior and experienced members of one of the construction professions, and for many years it was felt that they had a greater understanding of construction projects and the disputes that arise, than might be found in the courts. These days, however, the judges of the Technology and Construction Court have extensive experience of technical construction disputes. The high standards now evident in these courts are likely to be matched in practice by only a few arbitrators.

10.32   The court has powers to order that actions regarding related matters are joined (eg where disputes between an employer and contractor, and contractor and nominated sub-contractor, concern the same issues). This is much more difficult to achieve in arbitration. Even if all parties have agreed to the CIMA Rules, the appointing bodies must have been alerted and agreed to appoint the same arbitrator (Rules 2.6 and 2.7). If the same arbitrator is appointed, he or she may order concurrent hearings (Rule 3.7), but may only order consolidated proceedings with all the parties' consent (Rule 3.9), which is often difficult to obtain. The court's powers may therefore be an advantage in multi-party disputes, to avoid duplication of hearings and possible conflicting outcomes.

10.33   There remain, however, two key advantages to using arbitration. The first is that in arbitration the proceedings can be kept private – this is something which is usually of paramount importance to construction professionals and companies, and is often a deciding factor in selecting arbitration. In court, the proceedings are open to the public and the press, and the judgment is published and widely available.

10.34   The second advantage to the parties is that the arbitration process is consensual. The parties are free to agree on timing, place, representation and the individual arbitrator. This autonomy carries with it the benefits of increased convenience, and possibly savings in time and expense. The parties avoid long waiting lists currently running at the High Court, and choose a convenient time and place for the hearing. In arbitration, however, the parties have to pay the arbitrator and the cost of renting premises in which the hearing is held.

10.35   It should perhaps be noted that even where parties have selected arbitration under Article 9A, it is still open for them to elect litigation once a dispute develops. If, however, one party commences court proceedings, the other may ask the court to stay the proceedings on the grounds that an arbitration agreement already exists. This would not apply to litigation to enforce an adjudicator's decision, as Article 9A excludes all disputes regarding the enforcement of a decision of an adjudicator from the jurisdiction of the arbitrator.

Example of priced Activity Schedule

(see item 13)

A	Preliminaries – see breakdown	170,000.00
B	Demolition	40,000.00
C	Substructure including ground floor slab	178,000.00
D	Structural frame	265,000.00
E	Upper floor and staircase structures	112,000.00
F	Roof structure and coverings	58,000.00
G	External walls	203,000.00
H	Windows and external doors	102,000.00
J	Internal walls and doors therein	75,000.00
K	Partitioning and doors therein	45,000.00
L	Plasterwork	21,000.00
M	Screeds	33,000.00
N	Suspended ceilings	53,000.00
P	Wall tiling	16,000.00
Q	Floor tiling	11,000.00
R	Other floor finishings	64,000.00
S	Metalwork	30,000.00
T	Fittings and fixtures	35,000.00
U	Decorations	12,000.00
V	Sanitary installation and fittings	48,000.00
W	Rainwater installation	8,500.00
X	Mechanical services – see breakdown	297,000.00
Y	Electrical services – see breakdown	222,000.00
Z	Underground drainage	16,500.00
AA	External works	75,000.00
		£2,190,000.00

Note that the above excludes P.C. Sums and profit thereon, Provisional Sums and Provisional Quantities.

Appendix B - Amendment 12 to IFC98: Notices

4·2·3   This clause follows the terms of S.110(2). Leading Counsel advised that, as would frequently be the case, if the Employer intended to pay the amount stated as due in the certificate for interim payment it would be unnecessary to issue the S.110(2) and clause 4·2·3(a) notices which would add an unnecessary additional administrative burden on the Employer. Clause 4·2·3(c) therefore provides that where the Employer does not give the written notice under clause 4·2·3(a) he "shall pay the Contractor the amount due pursuant to clause 4·2(a)". If however the Employer proposes to withhold or deduct from any amount due pursuant to clause 4·2(a) he must comply with clause 4·2·3(b) (see below) and give the necessary written notice under clause 4·2·3(b).

4·2·3(b)   This clause is included in compliance with Ss.111(1) and (2). Once the contract on IFC 84 is in force the "five days" prescribed in this clause is an agreement to which S.111(3) refers. The position if, in the five days before the final date for payment, some defect were to appear in the Works, is not dealt with in the Section. It is suggested that in this situation the law of abatement will apply and the would-be paying party may be justifiably able to maintain that the amount of the proposed payment is not due in whole or in part so a 5-day notice is not required in order not to pay it. Whether this suggestion is correct will depend on whether the law of abatement is unaffected by the Act.

Practice Note 1

Construction Industry Scheme (CIS) – Inland Revenue

Income and Corporation Taxes Act 1988 (ICTA)
The Income Tax (Sub-Contractors in the Construction Industry) (Amendment)
Regulations 1998 S.I. No. 2622

1   The current statutory tax deduction scheme for which JCT Forms provide (e.g. JCT 98 clause 31) is to be replaced by the Construction Industry Scheme (CIS) **which starts on 1st August 1999.**

2   JCT has prepared new provisions headed "Construction Industry Scheme" to replace the existing provisions headed "Statutory tax deduction scheme" which will be invalid after 31st July 1999. The new provisions are set out in Amendments numbered "1: June 1999" to the 1998 editions of the following Forms:

JCT 98 All versions
WCD 98
IFC 98
Agreement for Minor Building Works 98
Measured Term Contract 98
Prime Cost Contract 98
Nominated Sub-Contract Conditions 98
Named Sub-Contract Conditions 98
Management Contract Conditions 98
Works Contract Conditions 98

3   For those **existing contracts** on JCT Forms under which payments will be made after 31st July 1999 a short Supplementary Agreement is set out at **Appendix 1**. Its effect is to substitute the new clause "Construction Industry Scheme (CIS)" for the previous clause "Statutory tax deduction scheme". **Alternatively** the parties may deal with the change of clause by an exchange of letters. The parties should check with IR14/15 (CIS) the status of the Employer under the CIS.

Outline of CIS

4   An essential feature of CIS is that a 'contractor' cannot make a payment under a contract for 'construction operations' to a 'sub-contractor' unless the 'sub-contractor' possesses either a tax certificate or certifying document or a registration card of whose validity the 'contractor' must be satisfied. If there is a valid tax certificate or certifying document the 'contractor' can pay an amount due without making the 'statutory deduction'; if there is a card the 'contractor' who is making the payment must take off from that payment the 'statutory deduction'.

contractor: this term is far wider than the normal meaning of the word and includes not only construction companies and firms but also many Employers including local authorities and Government Departments; and also "some businesses whose trading activities do not include construction operations . . . , but who may regularly carry out or commission construction work on their own premises or investment properties. Such a concern is a contractor if its average annual expenditure on construction operations in the period of three years ending with its last accounting date exceeds £1 million. Non-construction businesses which have not been trading for the whole of the last three years are also contractors if their total expenditure on construction operations for the part of that period of three years during which they were trading exceeds £3 million. In this way large manufacturing concerns, department stores, breweries, banks, oil companies, insurance companies and property investment companies may become contractors. Any business which becomes a contractor because one of the above conditions is met will continue to be one until it can satisfy the Inland Revenue that its expenditure on construction operations has been less than £1 million in each of three successive years." IR14/15 (CIS)

It is therefore vital under JCT Forms that they state e.g. in an Appendix entry whether the Employer is a 'contractor' for CIS purposes. Where the Employer is a 'contractor' then, again for CIS purposes, the Contractor will be a 'sub-contractor'.

"The following are NOT 'contractors'

• private householders having work done on their own premises (for example redecoration, repairs or an extension)

• a business where the trading activities do not include construction operations and the average annual expenditure on construction work on its own premises in recent years has been less than £1 million a year." IR14/15 (CIS)

sub-contractor: this term means "any business which has agreed to carry out construction operations for another business or public body which is a 'contractor ' – whether by doing the operations itself, or by having them done by its own employees or in any other way. Sub-contractors include concerns normally known as main contractors, where they are engaged by a client who is a contractor, for example a local authority." IR14/15 (CIS)

construction operations: the term is defined by ICTA S.567 which is reproduced at **Appendix 2**.

statutory deduction: this means the amount which a 'contractor' has to withhold from a payment to a 'sub-contractor' on account of tax and class 4 National Insurance Contributions, where appropriate. The rate of the deduction may be changed by Treasury Order. Any Tax Office will supply the current rate of deduction.

IFC 98   **Guidance Notes to Amendment 3: 2001**

Items 1-4   **Terrorism cover**

---

1. The insurance provisions in the JCT Main Contracts require the Employer or Contractor, as the case may be, to take out insurance against All Risks or specified perils. Each of these includes cover for loss or damage by fire or explosion howsoever caused. There is therefore no need to refer specifically to obtaining cover where the fire or explosion is caused by terrorism. The requirement already exists.

2. The provisions in Amendment 3 (which reiterate, in principle, those previously issued in TC/94/IFC) deal with the situation where terrorism cover, as defined in the contracts, is withdrawn or discontinued during the currency of the contract. In summary they provide that the Employer has the option of determining the employment of the Contractor or of requiring the Contractor to complete the Works at the Employer's cost.

3. Amendments to take account of these alternatives have also been issued for the Sub Contract NAM/SC.

4. 'Terrorism' for projects on the UK mainland has the narrow definition given in the Reinsurance (Acts of Terrorism) Act 1993 rather than the wider meaning set out in the Terrorism Act 2000.

   The 1993 Act describes it essentially as activities directed to the overthrow or influencing of the government, whereas the 2000 legislation is more extensive in that it covers violence for political, religious or ideological ends in order to influence the government or to intimidate the public.

5. The references applicable to Northern Ireland in the definition of All Risks Insurance in the relevant Forms have been removed. The JCT decided that these could be dealt with more appropriately in the adaptations of the Forms, issued by the Royal Society of Ulster Architects, for Works carried out in Northern Ireland.

6. Evidence of terrorism cover may be required to be produced by the Party obtaining the insurance even where that Party is a local authority.

7. Further details on obtaining terrorism cover, the cost, rating, exceptions and inclusions, and the position of the government as reinsurer of last resort, can be found in JCT Practice Note 3 (Series 2) 'Insurance - Terrorism Cover' which is available from the usual outlets.

Mediation on a Building Contract or Sub-Contract Dispute

1   Disputes can be resolved **either** by obtaining a court order (litigation) or an arbitration award or, if the Contract so provides, by adjudication; **or** by an agreement between the parties involved. Mediation is a way in which the negotiation of such an agreement may be facilitated. It involves the appointment by the parties of an independent person. The qualifications required from such person will depend on the nature of the dispute. He is not appointed to impose a solution but to try and steer the parties themselves towards a settlement.

2   While the use of a Mediation should always be considered it may not be suitable if one or more of the following factors are present: the dispute is one on which a legal precedent is being sought; either party needs an injunction to preserve the status quo pending a legal decision on the dispute; one or other party wants a public hearing; one or other party is not genuine in wanting to reach an agreed, as distinct from an imposed, settlement.

3   If the parties consider that a Mediator could assist in achieving a settlement it is advisable to draw up a simple agreement on the conduct of the Mediation: an example of such an agreement is annexed to this Note: **Example A**. The main matters which should be covered in such an agreement are:

–   a clear and precise statement of the issues in the dispute;

–   a declaration of the parties' wish to resolve the dispute with the help of a Mediator;

–   the period during which the Mediation is to take place but leaving either party free to withdraw without giving any reasons and so bringing the Mediation to an end; or for the parties to extend the period;

–   the name and qualifications of the Mediator;

–   a provision that the Mediation is to be conducted on a confidential and on a without prejudice basis unless and until, and to the extent that, the parties otherwise agree: and that no party will in any legal proceedings on the dispute or on any related matters call the Mediator to give evidence;

–   a date for meeting the Mediator to discuss *inter alia* the method by which the Mediation is to be conducted;

–   the question of how the costs of the Mediation are to be dealt with; that which is generally accepted (and is included in the annexed example of a Mediation Agreement) is that each party bears its own costs but with the fee of the Mediator being shared equally;

–   a provision that, if the Mediation results in a settlement of some or all of the issues, the parties will execute a binding agreement setting out the terms of the settlement.

4   The effect on the progress of the work of agreeing to a Mediation should be considered and agreement reached on how that effect is to be dealt with; and it may be that if a rapid, binding, decision is required, bearing in mind that Mediation may not necessarily end in an agreement between the parties, an arbitration under Rule 7 of the JCT Arbitration Rules (short procedure with a hearing) might be considered to be more appropriate. The application of Rule 7 does however require the agreement of both parties.

5   The agreement for Mediation does not prevent any party commencing or, if commenced, continuing arbitration, litigation or other proceedings.

6   It is recommended that, unless and until agreement is reached with the assistance of a Mediator, the parties should proceed with the contract as if there were no dispute.

7   If a dispute involves more than two contracting parties, that is more than the Employer and Contractor or the Contractor and a sub-contractor, the parties will have to adapt the arrangements suggested in this Practice Note which are intended for a dispute between two parties.

8   The Architect, Quantity Surveyor or other appointed consultant will usually need to be involved in the Mediation so that their comments are available to the parties.

9   If the settlement of a dispute between the Employer and the Contractor varies or negates a decision of the Architect or the Quantity Surveyor or other appointed consultant they should be given a copy by the Employer of the terms of settlement. Such terms might affect or be considered by them to affect their professional responsibilities under their agreements with the Employer.

10 Examples of the following documentation relevant to a Mediation are annexed to this Practice Note:

    A: Mediation Agreement;

    B: Agreement appointing a Mediator;

    C: Agreement following the resolution of a dispute after a Mediation.

11 Lists of Mediators are maintained by the Royal Institute of British Architects, the Royal Institution of Chartered Surveyors, the Association of Consulting Engineers, the Centre for Dispute Resolution and the Chartered Institute of Arbitrators.

12 The following is being included in JCT Forms as a footnote to the provisions setting out the arbitration agreement: "It is open to the Employer and the Contractor/to the Contractor and the sub-contractor to resolve disputes by the process of Mediation: see Practice Note 28 'Mediation on a Building Contract or Sub-Contract Dispute'."

13 This Practice Note is not applicable to Scotland. As and when the Scottish Building Contract Committee introduces mediation in Scotland it would make the necessary appointments.

References

Publications

JCT (1987) Practice Note 22: *Insurance*

JCT (1987) Practice Note 23: *A Contract Sum Analysis*

JCT (1995) Practice Note 27: *The Application of the Construction (Design and Management) Regulations 1994*

JCT (1995) Practice Note 28: *Mediation on a Building Contract or Sub-Contract Dispute*

JCT (1999) Practice Note 1: (series 2): *Construction Industry Scheme*

JCT (1999) Practice Note 2: (series 2): *Adjudication in JCT Forms*

JCT (2001) Practice Note 3: (series 2): *Insurance – Terrorism Cover*

JCT (forthcoming) Practice Note 6: (series 2): *Deciding on the Appropriate Form of JCT Main Contract*

Chappell, D and Powell-Smith, V (1999) *The JCT Intermediate Form of Contract.* Oxford, Blackwell Science Ltd

Cox, S and Clamp, H (1999) *Which Contract?* (2nd edn). London, RIBA Publications

Jones, N F and Baylis, S E (1999) Jones & Bergman's *JCT Intermediate Form of Contract* (3rd edn). Oxford, Blackwell Science Ltd

Latham, Sir M (1994) *Constructing the Team.* London, HMSO

Lupton, S (1997) *Architect's Guide to Arbitration.* London, RIBA Publications

Lupton, S (1998) *Architect's Guide to Adjudication.* London, RIBA Publications

May, A (1995) *Keating on Building Contracts.* London, Sweet & Maxwell

Cases

Alfred McAlpine Homes North Ltd v Property and Land Contractors Ltd (1995)76 BLR 59

Archivent Sales & Developments Ltd v Strathclyde Regional Council (1984) 27 BLR 98 (Court of Session, Outer House)

B Mullan & Sons Contractors v John Ross (1996) 86 BLR 1

Balfour Beatty Building Ltd v Chestermont Properties Ltd (1993) 62 BLR 1

British Telecommunications plc v James Thompson & Sons (Engineers) Ltd (1996) 82 BLR 1 (Court of Session, Inner House)

Colbart Ltd v H Kumar (1992) 59 BLR 89

Croudace Ltd v London Borough of Lambeth (1986) 33 BLR 25 (CA)

Crown Estate Commissioners v John Mowlem & Co Ltd. (1994) 70 BLR 1 (CA)

Dawber Williams Roofing Ltd v Humberside County Council (1979) 14 BLR 70

Department of Environment for Northern Ireland v Farrans (Construction) Ltd (1981) 19 BLR 1 (NI)

FG Minter Ltd v Welsh Health Technical Services Organisation (1980) 13 BLR 1 (CA)

Greater London Council v Cleveland Bridge and Engineering Co Ltd (1986) 34
  BLR 50 (CA)
Glenlion Construction Ltd v The Guinness Trust (1987) 39 BLR 89
H Fairweather & Co Ltd v London Borough of Wandsworth (1987) 39 BLR 106
H W Nevill (Sunblest) Ltd v William Press & Son Ltd (1981) 20 BLR 78
*Holland Hannen & Cubitts (Northern) Ltd v Welsh Health Technical Services
  Organisation* (1985) 35 BLR 1 (CA)
J F Finnegan Ltd v Ford Sellar Morris Developments Ltd (1991) 53 BLR 38
*Kensington and Chelsea and Westminster Health Authority v Wettern
  Composites* (1984) 31 BLR 57
Kruger Tissue (Industrial) Ltdv Frank Galliers Ltd (1998) 57 ConLR 1
Leedsford Ltd v The Lord Mayor, Alderman and Citizens of the City of Bradford
  (1956) 24 BLR 45 (CA)
London Borough of Houndslow v Twickenham Garden Developments (1970) 7
  BLR 81
London Borough of Merton v Stanley Hugh Leach Ltd (1985) 32 BLR 51
  (Chancery Division)
*Lubenham Fidelities and Investments Co Ltd v South Pembrokeshire District
  Council* (1986) 33 BLR 39 (CA)
Mac-Jordan Construction Ltd v Brookmount Erostin Ltd (1991) 56 BLR 1
Michael Saliss & Co Ltd v Calil and William F Newman & Associates (1987) 13
  ConLR 69
Pacific Associates Inc v Baxter (1987)13 ConLR 80
Peak Construction (Liverpool) Ltd v McKinney Foundations Ltd (1970) 1 BLR
  111 (CA)
Penwith District Council v VP Developments Ltd, 21 May 1999, unreported
Percy Bilton Ltd v Greater London Council (1981)17 BLR 1 (HL)
R Burden Ltd v Swansea Corporation [1957] 3 All ER 243 (HL)
Rayack Construction Ltd v Lampeter Meat Co Ltd (1979) 12 BLR 30
Rotherham MBC v Frank Haslam Milan and MJ Gleeson (1996) 78 BLR 1 (CA)
Ruxley Electronics and Construction Ltd v Forsyth (1995) 73 BLR 1
Scottish Special Housing Association v Wimpey Construction UK Ltd (1986) 34
  BLR 1
Sutcliffe v Chippendale & Edmondson (1971) 18 BLR 149
Sutcliffe v Thackrah (1974) 4 BLR 16 (CA)
Temloc v Errill Properties (1988) 39 BLR 30 (CA)
Townsend v Stone Toms & Partners (1984) 27 BLR 26 (CA)
Wates Construction (London) Ltd v Franthorm Property Ltd (1991) 53 BLR 23 (CA)
Wates Construction (South) Ltd v Bredero Fleet Ltd (1963) 63 BLR 128
West Faulkner Associates v London Borough of Newman (1992) 61 BLR 81
Whittal Builders Co Ltd v Chester-le-Street District Council (1987) 40 BLR 82

Subject index *(by paragraph number)*